Cambridge Tracts in Mathematics
and Mathematical Physics

GENERAL EDITORS
H. BASS, J. F. C. KINGMAN, F. SMITHIES,
J. A. TODD AND C. T. C. WALL

No. 61

GENERALIZED CLIFFORD PARALLELISM

GENERALIZED CLIFFORD PARALLELISM

J. A. TYRRELL
Reader in Mathematics,
King's College London

AND

J. G. SEMPLE
Professor Emeritus of Mathematics,
King's College London

CAMBRIDGE
AT THE UNIVERSITY PRESS
1971

CAMBRIDGE UNIVERSITY PRESS
Cambridge, New York, Melbourne, Madrid, Cape Town, Singapore, São Paulo, Delhi

Cambridge University Press
The Edinburgh Building, Cambridge CB2 8RU, UK

Published in the United States of America by Cambridge University Press, New York

www.cambridge.org
Information on this title: www.cambridge.org/9780521080422

© Cambridge University Press 1971

This publication is in copyright. Subject to statutory exception
and to the provisions of relevant collective licensing agreements,
no reproduction of any part may take place without the written
permission of Cambridge University Press.

First published 1971
This digitally printed version 2008

A catalogue record for this publication is available from the British Library

Library of Congress Catalogue Card Number: 74–134625

ISBN 978-0-521-08042-2 hardback
ISBN 978-0-521-09184-8 paperback

CONTENTS

CONTENTS

PREFACE

The subject of parallelism in elliptic non-Euclidean space is one that has hitherto received little attention in works on geometry. In fact, although Clifford's original theory of parallel lines in three-dimensional elliptic space goes back nearly a hundred years, the generalization to higher space, due to Y.-C. Wong, was not devised until comparatively recently. On reading Wong's work (which is predominantly algebraic and mainly confined to geometry over the real field) it occurred to the authors that the subject could more naturally be developed in the wider context of complex projective geometry and the present tract was conceived with this object in mind.

The general plan of the book, together with some points of history and motivation, is the subject of Chapter 1 and need not be anticipated here. As to the style of writing, we have tried to develop the earlier Chapters in a careful and rigorous manner, paying full attention to the details of the arguments; in the later Chapters, however, we have felt free to use the less formal language of the classical geometers and to omit some of the more routine proofs and calculations.

We wish to thank Dr D. Kirby of the University of Southampton for the meticulous care with which he read our page proofs and for his numerous corrections and emendations; and our warm thanks are also due to the officers of the Cambridge University Press for their unfailing courtesy and assistance in the preparation of this volume.

King's College London   J.A.T.
June 1970   J.G.S.

CHAPTER 1

INTRODUCTION

1. Parallelism in the plane

When the study of plane Euclidean geometry is being developed by coordinate methods, there are two normally accepted procedures for dealing with the phenomenon of parallelism. In the first of these procedures, we regard the Euclidean plane, effectively, as a real projective plane with one line removed—a line which we denote by l_∞; we can then regard the projective plane, conversely, as an *extended Euclidean plane*; and thereafter we define parallel lines as lines whose closures in the extended Euclidean plane meet on l_∞. By the second procedure we first introduce a distance function and the concept of perpendicularity, showing that the shortest distance from a point to a line is the perpendicular distance; and we then say that two lines are parallel if each point of either is at the same perpendicular distance from the other.

When we come to study non-Euclidean geometry, it is natural to investigate the possibility of generalizing one or other of the above approaches to parallelism. A non-Euclidean plane, as we may recall, is basically a real projective plane in which there is singled out a particular polarity ω—the *absolute polarity*; that is to say, in a chosen system of homogeneous coordinates (x_0, x_1, x_2) in the plane, there is singled out a bilinear relation

$$\sum_{i=0}^{2} \sum_{j=0}^{2} a_{ij} x_i x_j' = 0 \qquad (1.1)$$

such that the matrix $A = (a_{ij})$ is real, non-singular and symmetric; and two points (x_0, x_1, x_2) and (x_0', x_1', x_2') are said to be *conjugate* with respect to the polarity ω if their coordinates satisfy (1.1). The points of the plane that are conjugate to themselves are those of the (non-singular) *absolute conic* Ω whose equation is

$$\sum_{i=0}^{2} \sum_{j=0}^{2} a_{ij} x_i x_j = 0;$$

and the non-Euclidean plane is said to be *hyperbolic* or *elliptic* according as Ω does or does not contain any real points; or, as we usually say, according as Ω is a *real* or a *virtual*† conic.

In the hyperbolic case Ω disconnects the projective plane into two regions, one homeomorphic with a disk and the other with a Möbius strip; but it is customary to think of the hyperbolic plane as properly consisting only of the points of the former region, those of the latter region being called *ultra-infinite points* and those of Ω *points at infinity*. The elliptic plane, on the other hand, consists of *all* the points of the underlying projective plane.

The *non-Euclidean distance* $d(P, Q)$ between points P and Q of a non-Euclidean plane is defined in terms of the cross-ratio $\{P, Q; Q', P'\}$, where P' and Q' are the points of the line PQ that are conjugate to P and Q respectively in the polarity ω. The usual definition is

$$\left.\begin{array}{c}\cosh^2(d(P, Q)) \\ \cos^2(d(P, Q))\end{array}\right\} = \{P, Q; Q', P'\},$$

the hyperbolic or trigonometric cosine being taken according as the plane is hyperbolic or elliptic.‡ The use of the term 'point at infinity' for a point of Ω (in the hyperbolic case) is justified by the fact that $d(P, Q)$ tends to infinity if P is kept fixed while Q is made to approach a point of Ω.

After these preliminaries we may now consider possible generalizations to non-Euclidean geometry of the two ways of introducing parallel lines in the Euclidean plane. By analogy with the first of these ways we may define two lines of a non-Euclidean plane to be parallel if they meet 'at infinity', i.e. on Ω. In the hyperbolic case this definition makes sense, and its consequences have been fully worked out in treatises on non-Euclidean geometry (cf., for example, Coxeter [6] which contains an extensive bibliography). In the elliptic case, however, the definition is empty because Ω contains no real points.

† Since virtual conics contain no real points, they do not exist (as loci) in the same sense as real conics. They are associated, however, with well-defined real polarities and thereby acquire a conventional 'existence' as conics in the real projective plane.

‡ The formulae given above do not define $d(P, Q)$ unambiguously in either case. For the conventions necessary to their interpretation we refer the reader to the account given in Coxeter [6].

As regards the possible generalization of the 'equidistance' definition of parallel lines, we note first that *perpendicular lines* in either kind of non-Euclidean geometry are defined to be lines which are conjugate with respect to the absolute polarity ω, and that the shortest distance from a point to a line is the perpendicular distance. We then find, however, that no two distinct lines can ever have the 'equidistance' property; and, in fact, the locus of a point which is at a fixed perpendicular distance from a given line – *an equidistant curve* – is (algebraically) a conic touching Ω at its intersections with the line. There is, accordingly, no analogue, in either kind of non-Euclidean plane, of the equidistance definition of parallelism in the Euclidean plane.

2. Parallel lines in 3-dimensional space

Consider now the corresponding problems in 3-dimensional space. Parallel lines in Euclidean 3-dimensional space can be introduced in either of two ways, either as lines that meet on a conveniently introduced 'plane at infinity' or as lines that are everywhere equidistant; and we look again at the possibility of adapting one or other of these definitions to the circumstances of 3-dimensional non-Euclidean geometry.

A 3-dimensional non-Euclidean space is a 3-dimensional real projective space in which there is singled out a certain fixed (non-singular) quadric polarity ω—the *absolute polarity*—given in a chosen coordinate system by an equation

$$\sum_{i=0}^{3} \sum_{j=0}^{3} a_{ij} x_i x_j' = 0,$$

where $A = (a_{ij})$ is a non-singular real symmetric matrix. We again distinguish different kinds of non-Euclidean space according to the nature of the non-singular *absolute quadric* Ω whose equation is

$$\sum_{i=0}^{3} \sum_{j=0}^{3} a_{ij} x_i x_j = 0.$$

Specifically, the space is *elliptic* if Ω is a virtual quadric (containing no real points) and *hyperbolic* if Ω is a real quadric of the kind that is homeomorphic with an ordinary 2-dimensional sphere; and in the latter case we again distinguish (as in the plane) between ordinary, infinite and ultra-infinite points.† *Non-Euclidean distance* is defined, in either kind of space, by the same formulae as in the plane; and two lines p and q are defined to be *perpendicular* if they are conjugate with respect to ω, i.e. if each meets the polar line of the other. It is easy to verify, then, that the shortest distance from a point to a line is the perpendicular distance.

As regards the possible definitions of parallel lines, the first possibility is to define two lines as being parallel if they meet on Ω; and, just as in the plane, this definition makes sense in the hyperbolic but not in the elliptic case. With the equidistance definition, however, an exciting new possibility arises, of which there was no analogue in the plane. The locus ϕ of points at a fixed perpendicular distance from a given line l – an *equidistant surface* – turns out to be a quadric surface which (algebraically) touches Ω at its two intersections with l; and there is now the possibility that this quadric ϕ may be of the type that contains real generating lines. It then turns out that this possibility is realized if (and only if) the non-Euclidean space is elliptic. To recapitulate formally:

In 3-dimensional elliptic space, the locus of points at a fixed perpendicular distance from a given line l is a real quadric surface ϕ with (two systems of) real generating lines.

Any line of either generating system on ϕ is therefore parallel to l in the 'equidistance' sense, and it is said to be *Clifford parallel*‡ to l. Plainly, through any general point P (of elliptic space) there pass two lines that are Clifford parallel to a given line l, one generator of each system on the quadric such as ϕ that contains P; and it is possible to introduce conventions whereby one of these lines can be said to be *left-parallel* to l and the other *right-parallel* to l. Left-parallelism and right-parallelism each turn out to be

---

† There is, of course, another kind of real non-singular quadric in real projective space, namely the kind – homeomorphic with a torus – which possesses two systems of real generating lines; but for various reasons that need not detain us here, this kind of quadric does not give rise to an acceptable kind of non-Euclidean geometry.

‡ In the older literature, Clifford parallel lines were sometimes called *paratactics*.

transitive relations, though Clifford parallelism without qualification is not. An important consequence of the above remarks is that there exist *fibrations* of elliptic 3-space by systems of mutually left- (or right-) parallel lines, where by a fibration we mean a system of which exactly one line passes through each point of space (with no exceptions).

3. Clifford's original definition of parallelism

We now refer to another way of approaching the concept of Clifford parallelism, more in the spirit of Clifford's original discovery (cf. Clifford [5]). Suppose that we are given a point P and a line q in elliptic 3-space, and let q' be the polar line of q in the absolute polarity ω. Then the transversal line from P to q and q' is the perpendicular from P to q, and the shortest distance from P to q is measured along this perpendicular. Now suppose that P moves along a line p whose polar line is p'. It can be shown then that the distance from P to q varies continuously as P moves along p, and it attains its maximum and minimum values when P lies on one of the two transversal lines t and t' that can be drawn to meet p, p', q and q'. These lines t and t' are the common perpendicular transversals of p and q; and, provided p and q are in sufficiently general position, there are no others. But, as Clifford noticed, there is a case of exception in which p, p', q and q' belong to the same regulus of generators of a ruled quadric R. In this case p and q possess an infinite number of common perpendicular transversals – the lines of the regulus on R complementary to that containing p, q, p' and q' – and the distances between p and q along these transversals are all equal. These observations lead to the following projective characterization of Clifford parallel lines:

Clifford's definition of parallelism: *Two lines p and q are (Clifford) parallel if (and only if) they and their polar lines with respect to ω are four generators of one regulus on a ruled quadric R.*

A quadric such as R, i.e. one which contains two Clifford parallel lines and their polars, possesses the following two (characteristic) properties:

(*a*) Any two non-intersecting generators of R are Clifford parallel; more precisely, the generators of one regulus on R are all mutually left-parallel, while those of the other are mutually right-parallel; and

(*b*) R is autopolar with respect to ω, i.e. it contains the polar line of every one of its generators.

A quadric with properties (*a*) and (*b*) will be called a *Clifford quadric*,† and each of its two reguli will be called a *Clifford regulus*. In the higher dimensional analogues of Clifford parallelism which we shall be studying, it will be found that the appropriate analogue of a Clifford regulus plays a fundamental role.

4. Isoclinic planes in Euclidean space E_4

Further insight into Clifford parallelism may be gained by considering the problem, in 4-dimensional Euclidean space E_4, of defining the angle between two planes (2-dimensional subspaces of E_4).

If π_1 and π_2 are two non-orthogonal planes through a point O of E_4, then one may consider a variable vector v through O in π_1 and calculate the angle between v and its orthogonal projection on π_2. This angle, as it turns out, varies with v in general, achieving its maximum and minimum values when v lies in one or other of two mutually perpendicular directions in π_1. Thus from this approach it is not clear how to define the angle between π_1 and π_2. (For the answers to this and related questions the reader may consult Forsyth[10].) There is a case of exception, however, that can and does occur, in which the angle θ between v and its projection on π_2 is independent of the choice of v in π_1. Here there is no doubt as to how to define the angle between π_1 and π_2; and we say that π_1 and π_2 are *isoclinic* at an angle θ.

It is not our purpose to investigate the properties of isoclinic planes in any detail; this has been done elsewhere (see, for example, Forsyth[10] and Manning[19]). We only wish to point out that Euclidean space E_4 may be thought of as acquiring its Euclidean structure in terms of a 'solid at infinity' together with

† We should point out that Coxeter[6] uses the term 'Clifford surface' to describe any quadric which possesses property (*a*) but not necessarily (*b*).

a virtual quadric Ω in this solid; and the solid at infinity has then the structure of an elliptic 3-dimensional space. It may now be verified easily that *two planes of E_4 through O are isoclinic if and only if they meet the elliptic 3-dimensional space at infinity in Clifford parallel lines*.

To mention only one consequence of this, consider a fibration of the solid at infinity by a system of mutually left-parallel lines, say; and join these by planes to a fixed origin O in E_4. Exactly one of these planes will pass through any point of E_4 other than O. Thus if we consider a 3-dimensional sphere S^3 in E_4 with centre O, we see that the planes in question will cut S^3 in a family of non-intersecting great circles such that exactly one passes through any point of S^3. The decomposition space of this family of circles turns out to be a 2-dimensional sphere S^2; and what we have, in effect, is the well-known Hopf fibering of S^3 by 1-spheres over S^2. In terms of this construction one can further establish the existence of three mutually orthogonal fields of unit tangent vectors on S_3, thus demonstrating its parallelizability.

5. Generalization of Clifford parallelism

A $(2n-1)$-dimensional elliptic space EL_{2n-1} is a $(2n-1)$-dimensional real projective space in which a certain virtual quadric is assigned the role of absolute quadric Ω and the associated real non-singular polarity is the absolute polarity ω. An $(n-1)$-dimensional subspace Π of EL_{2n-1} has a polar space Π' with respect to ω, also of dimension $n-1$, and a line is said to be perpendicular to Π if it meets Π'. If we define distance in the same way as for the elliptic plane, we find that the shortest distance to Π from a point is the perpendicular distance, and we may make the following definition:

Two $(n-1)$-dimensional subspaces Π and Π_1 of EL_{2n-1} are to be called *Clifford parallel* if the perpendicular distance to Π_1 from any point P of Π is independent of the choice of P in Π.

This is no more than a straightforward generalization of the original concept of Clifford parallelism, and it is therefore remarkable that it was not systematically studied and developed until recently. The recent development, due to Wong[27], is coupled with a parallel study of isoclinic n-dimensional subspaces of Euclidean space E_{2n}, the connection with Clifford parallelism being analogous to that which we have described above (§4) for $n = 2$. Wong's development is primarily directed to the discovery of systems – more particularly maximal systems – of mutually Clifford parallel $[n-1]$'s† in EL_{2n-1}; and we may summarize his findings briefly as follows:

(i) For all values of n (> 1) there exist 1-dimensional systems of mutually Clifford parallel $[n-1]$'s in EL_{2n-1}. Important among such systems are those which Wong calls *additive sets*, and these have properties very similar to those of Clifford reguli in the case $n = 2$.

(ii) For given n there exists an r-dimensional system of mutually Clifford parallel $[n-1]$'s in EL_{2n-1} if and only if there exists a set of $r-1$ skew-symmetric real orthogonal matrices of order n which anticommute by pairs. The algebraic problem so arising is that of solving the well-known Hurwitz–Radon matrix equations, and the pairs of values (n, r) for which a solution exists are known‡ (cf. the Appendix to this book).

(iii) Except when $n = 2, 4$ or 8, there exist no n-dimensional systems of mutually Clifford parallel $[n-1]$'s in EL_{2n-1}, and consequently no fibrations of EL_{2n-1} based on such systems; but when $n = 2, 4$ or 8, such fibrations (or *space-filling systems*) do exist. The fibrations for $n = 4$ and $n = 8$ correspond, by a construction similar to that described above for $n = 2$, to the Hopf fiberings of the 7-sphere S^7 by S^3's over S^4 and of the 15-sphere S^{15} by S^7's over S^8.

Since Wong's treatment of Clifford parallelism, and the formulation of his results, were predominantly algebraic and limited to geometry over the real field, it seemed to the present authors that the subject invited an alternative development

---

† The symbol $[r]$ will be used in this book to denote an r-dimensional (flat) space.

‡ When $n = 3$, the maximum possible value of r is 1; so that in this, the next simplest case of Clifford parallelism after the classical case, the phenomenon is relatively uninteresting. This may partly explain why the problem of generalizing Clifford parallelism did not attract attention for such a long time.

within the general framework of complex projective geometry; and, in particular, it seemed highly desirable to seek some insight into the geometric construction and properties of systems of mutually Clifford parallel spaces. In this book we give such a development, essentially independent of that given by Wong, and consisting basically of a chapter in the projective geometry of complex projective space S_{2n-1} relative to a non-singular quadric primal Ω_{2n-2}. After some preliminaries (Chapter 2), we introduce the notion of Clifford parallel spaces with respect to Ω, using a straightforward generalization of Clifford's own definition (§ 3), and we rediscover Wong's additive sets (which we call Clifford reguli) and develop their properties (Chapter 3). This is followed by a mainly algebraic exposition of what we call *linear* systems of Clifford parallels (Chapter 4), analogous to, but rather more general than the systems discovered by Wong. The next Chapter is devoted to geometrical constructions for systems of Clifford parallels, particular attention being paid to the two most interesting cases, $n = 4$ and $n = 8$. The remainder of the book (Chapters 6 and 7) is concerned with representation theory; and this includes in particular a detailed account of a new sequence of algebraic varieties – here called T-models – which serve as birational maps of a specially significant kind of the pairs of polar $[n-1]$'s with respect to a quadric Ω in S_{2n-1}. Among the applications of our representation theory we encounter in Chapter 7 an apparently new construction for, and extended interpretation of, that well-known geometrical curiosity, the Study triality correspondence between the points and the two types of generating solids of a quadric Ω in S_7.

PRELIMINARIES OF GEOMETRY IN S_{2n-1}

In this book we shall be concerned throughout, except when the contrary is stated, with geometry in a complex $(2n-1)$-dimensional projective space S_{2n-1} relative to a non-singular 'absolute' quadric primal Ω. We devote the present chapter, therefore, to recording some preliminary material about the appropriate coordinate systems, the properties of a quadric Ω, the definition and properties of the well-known type of algebraic variety which we call a *regulus* of $(n-1)$-dimensional spaces of S_{2n-1}, and some definitions and details relating to Grassmannian and Veronesean varieties.

1. Coordinate system in S_{2n-1}

In S_{2n-1} we shall generally use a system of allowable homogeneous coordinates

$$(x_0, ..., x_{n-1}, y_0, ..., y_{n-1}) \tag{1.1}$$

based on a pair of skew $(n-1)$-dimensional reference spaces X, Y such that X contains the reference points $X_0, ..., X_{n-1}$ while Y contains $Y_0, ..., Y_{n-1}$. We shall write x (resp. y) for the column vector of the x_i (resp. y_i) and shall refer to (1.1) as the point (x, y). We note then that

(i) any $[n-1]$ of S_{2n-1} which is skew to the reference space Y has a matrix equation of the form

$$y = Ax, \tag{1.2}$$

where A is a uniquely defined $n \times n$ matrix of constants, and

(ii) the same $[n-1]$ is skew also to X if and only if $|A| \neq 0$. Two $[n-1]$'s given by $y = A_1 x$ and $y = A_2 x$ are skew to one another if and only if $|A_1 - A_2| \neq 0$.

2. The non-singular quadric Ω

We shall generally choose the coordinate system in such a way that the equation of Ω reduces to one or other of two standard forms which we now consider separately.

The first standard form $x^T x + y^T y = 0$. For this form, in addition to a proper choice of the unit point, the simplex of reference must be self-polar for Ω; and, in particular, the reference spaces X and Y must be polars of each other for Ω. Conversely, moreover, it should be noted – as may easily be verified – that any pair of skew mutually polar $[n-1]$'s may be chosen as the reference spaces X, Y of a coordinate system in which Ω has the equation $x^T x + y^T y = 0$. The condition that a pair of polar $[n-1]$'s should be skew means that neither of them touches Ω, i.e. meets Ω in a singular quadric.

When Ω has the equation $x^T x + y^T y = 0$ and Π is the $[n-1]$ with equation $y = Ax$, then it is easily seen that

(i) the section of Ω by Π is non-singular if and only if $|I + A^T A| \neq 0$, where I is the $n \times n$ unit matrix,

(ii) if Π does not meet X, so that $|A| \neq 0$, then the polar space Π' of Π has the matrix equation $y = -A^{-1T}x$, and

(iii) Π lies on Ω (i.e. is self-polar) if and only if $A^T A = -I$.

The second standard form $x^T y = 0$. For this form, besides the choice of unit point, the reference spaces X and Y must be a pair of skew generating spaces of Ω; and, conversely, any such pair of skew generators of Ω can be taken to be the reference spaces X, Y of a coordinate system in which Ω has the equation $x^T y = 0$.

When Ω has the equation $x^T y = 0$ and Π is the $[n-1]$ given by $y = Ax$, then

(i) the section of Ω by Π is non-singular if and only if

$$|A + A^T| \neq 0,$$

(ii) the polar space Π' of Π is given by $y = -A^T x$, and

(iii) Π is a generator of Ω if and only if A is skew-symmetric.

The equation $x^T y = 0$ can also be interpreted as that of a correlation between the spaces X and Y with the property that points P and Q in X and Y respectively correspond in the correlation if and only if their join PQ lies on Ω. Thus Ω induces in this way what we may call a *natural* correlation κ between any skew pair of its generating spaces.

We may recall here the following well-known properties of generators of Ω:

(*a*) Ω possesses two distinct systems of $(n-1)$-dimensional generating spaces (which we shall call the α-spaces and the β-spaces), each system being of dimension $\frac{1}{2}n(n-1)$.

(*b*) If n is even, two generators of the same system do not meet in general but may meet in a space of an odd number of dimensions, while two generators of opposite systems meet always in a point or in a space of an even number of dimensions.

(*c*) If n is odd, two generators of the same system meet either in a point or in a space of an even number of dimensions, while two of opposite systems do not in general meet but may meet in a space of an odd number of dimensions.

3. Reguli in S_{2n-1}

In the developments to follow we shall be much concerned with the type of algebraic variety R of S_{2n-1} that is generated by the lines joining pairs of points which correspond in a non-singular collineation between two skew $[n-1]$'s. Any such R is readily identified as a Segre product of a line by an $(n-1)$-dimensional space, and it is known, as such, to be a non-singular variety R_n^n normal in S_{2n-1}. The product structure shows that R is simply generated not only by its ∞^{n-1} lines, which are called its *directrices*, but also by ∞^1 mutually skew $[n-1]$'s – unisecant to the directrices – which are called its *generators*. A variety such as R, particularly when it is regarded as the aggregate of its generators, will be called a *regulus* (of $[n-1]$'s of S_{2n-1}). There exist, of course, specialized and degenerate types of the aggregate in question; but we shall find it convenient throughout this book to reserve the term 'regulus of $[n-1]$'s' to mean always an aggregate which conforms precisely to the general definition we have given.

Analytically, if we define R by means of the collineation

$$\rho y = Ax \quad (|A| \neq 0)$$

between the reference spaces X and Y, we find that the generators of R are given, for different values† of the scalar parameter λ, by the equation

$$y = \lambda A x.$$

With no loss of generality (i.e. by a transformation of the co-ordinates x_i) we can take A to be the $n \times n$ unit matrix, in which case the generators of R are given by the equation

$$y = \lambda x,$$

and the equations of R are

$$\frac{x_0}{y_0} = \frac{x_1}{y_1} = \ldots = \frac{x_{n-1}}{y_{n-1}}.$$

We note, in particular, that R is completely defined as an intersection of quadrics.

Among the elementary results about reguli we record briefly the following.

PROPOSITION 3.1. *The directrix lines of a regulus R cut collinear fields on its generators.*

This means, in effect, that the points of any set S_1 in a generator Π_1 are carried along the directrices passing through them into the points of a collinear set S_2 in any other generator Π_2. We shall express this connection frequently by saying that S_2 *lies over* S_1 (*or S_1 over S_2*) in R.

PROPOSITION 3.2. *The transversal lines of three mutually skew $[n-1]$'s of S_{2n-1} are the directrices of a regulus which contains the three $[n-1]$'s as generators.*

For the correspondence between two of the $[n-1]$'s determined by transversals drawn to them from points of the third is a collineation.

COROLLARY 3.2.1. *There is a unique regulus R containing three mutually skew $[n-1]$'s of S_{2n-1}.*

---

† The space Y corresponds to the improper value ∞ of λ. For any letter denoting a parameter, it will usually be clear from the context whether ∞ is to be included in its range of values.

PROPOSITION 3.3. *There exist* ∞^{2n-4} *solids (3-dimensional flat spaces) each of which meets every generator of a given regulus R in a line (these lines all lying over any one of themselves and generating a quadric surface); and precisely one of these solids passes through any point of S_{2n-1} that does not lie on R.*

The solids in question will be referred to as the *quadric-solids* of R.

We leave the proof of Prop. 3.3 as an exercise for the reader.

PROPOSITION 3.4. *Let S_{2p-1} and S_{2q-1} $(p+q=n)$ be two skew spaces which span S_{2n-1}; and let R_1 be a regulus of $[p-1]$'s of S_{2p-1}, while R_2 is a regulus of $[q-1]$'s of S_{2q-1}. Further, let there be given a homographic correspondence between the generators of R_1 and those of R_2. Then the $[n-1]$'s which join corresponding generators of R_1 and R_2 are the generators of a regulus of $[n-1]$'s in S_{2n-1}.*

Here again the proof may be left to the reader. A similar result holds for three or more homographically related reguli in mutually skew spaces which span S_{2n-1}.

4. The matrix cross-ratio theorem

We now establish a basic algebraic criterion for four $[n-1]$'s of S_{2n-1} to belong to a regulus.

THEOREM 4.1. *If $A_1, ..., A_4$ are four $n \times n$ matrices for which all the differences $A_i - A_j$ $(i \neq j)$ are non-singular, and if Π_i is the $[n-1]$ of S_{2n-1} with equation $y = A_i x$ $(i = 1, ..., 4)$, then $\Pi_1, ..., \Pi_4$ belong to a regulus R if and only if there exists a scalar c $(\neq 0, 1)$ such that*

$$(A_1 - A_3)(A_1 - A_4)^{-1}(A_2 - A_4)(A_2 - A_3)^{-1} = cI, \qquad (4.1)$$

where I is the $n \times n$ unit matrix. If the regulus R is thought of abstractly as a projective line whose points are the generators of R, then c is the cross-ratio $\{\Pi_1, \Pi_2; \Pi_3, \Pi_4\}$.

Proof. The non-singularity of the matrices $A_i - A_j$ $(i \neq j)$ means simply that the four spaces $\Pi_1, ..., \Pi_4$ are mutually skew.

Consider an arbitrary regulus R containing Π_1 and Π_2. Then R is defined by some non-singular collineation between the points $(\xi_1, A_1\xi_1)$ of Π_1 and the points $(\xi_2, A_2\xi_2)$ of Π_2, say

$$\rho\xi_2 = T\xi_1 \quad (|T| \neq 0),$$

and the general point of the variety R is of the form

$$(\xi_1 - \lambda T\xi_1, A_1\xi_1 - \lambda A_2 T\xi_1).$$

The directrices of R are obtained by fixing ξ_1 and allowing λ to vary, while the generators arise from fixing λ and allowing ξ_1 to vary. We therefore get the general generator by eliminating the vector ξ_1 between the two equations

$$x = \xi_1 - \lambda T\xi_1 \quad \text{and} \quad y = A_1\xi_1 - \lambda A_2 T\xi_1.$$

This yields $y = (A_1 - \lambda A_2 T)(I - \lambda T)^{-1}x,$

and we note that $\lambda = 0$ gives Π_1, while $\lambda = \infty$ gives Π_2.

If we now require Π_3 and Π_4 to belong to R, we must have

$$A_i = (A_1 - \lambda_i A_2 T)(I - \lambda_i T)^{-1} \quad (i = 3, 4) \tag{4.2}$$

for suitable scalars λ_3 and λ_4; and since $\Pi_1, \ldots, \Pi_4$ are distinct, λ_3 and λ_4 must be distinct and neither of them equal to 0 or ∞. A rearrangement of (4.2) gives

$$T = \lambda_i^{-1}(A_2 - A_i)^{-1}(A_1 - A_i) \quad (i = 3, 4)$$

so that, equating these two expressions for T, we have

$$\lambda_3^{-1}(A_2 - A_3)^{-1}(A_1 - A_3) = \lambda_4^{-1}(A_2 - A_4)^{-1}(A_1 - A_4).$$

This equation on rearrangement gives (4.1) with $c = \lambda_3/\lambda_4$, and since λ_3 and λ_4 are distinct non-zero constants we have $c \neq 0, 1$. Since the algebra is reversible, the first part of the theorem is thereby proved.

Further, since $\Pi_1, \ldots, \Pi_4$ are the members of R given respectively by the values $0, \infty, \lambda_3, \lambda_4$ of λ, we have

$$\{\Pi_1, \Pi_2; \Pi_3, \Pi_4\} = \{0, \infty; \lambda_3, \lambda_4\} = \lambda_3/\lambda_4 = c;$$

and this completes the proof of the theorem.

It is natural to denote the left-hand side of (4.1) by

$$\{A_1, A_2; A_3, A_4\}$$

and to call it the *cross-ratio of the four matrices*. It is then implicit in Theorem 4.1 that if $\{A_1, A_2; A_3, A_4\} = cI$, then – under permutations of $A_1, \ldots, A_4$ – the cross-ratio behaves like the cross-ratio of four numbers. For example:

$$\{A_2, A_1; A_3, A_4\} = c^{-1}I$$

and $$\{A_1, A_3; A_2, A_4\} = (1-c)I.$$

Formal algebraic proofs of these results involve some skilful matrix manipulations.

Ex. 1. Let

$$A_0 = \begin{pmatrix} \alpha_0 & \beta_0 \\ \gamma_0 & \delta_0 \end{pmatrix} \quad \text{and} \quad A_1 = \begin{pmatrix} \alpha_1 & \beta_1 \\ \gamma_1 & \delta_1 \end{pmatrix}$$

be two given matrices of order two with the property that the difference between any two of A_0, A_0^T, A_1, A_1^T is non-singular. For a general value of the scalar λ, show that the equation

$$\{A_1, A_0; A_0^T, X\} = \lambda I$$

has a unique solution for X, namely,

$$X = A(\lambda) \equiv (\lambda A_1 K - A_0)(\lambda K - I)^{-1},$$

where $K = (A_1 - A_0^T)^{-1}(A_0 - A_0^T)$. $A(\lambda)$ is equal to A_0, A_1 and A_0^T respectively when $\lambda = 0, \infty$ and 1. Show also that $A(\lambda)$ is equal to A_1^T when λ takes the value λ_1 given by

$$(\beta_0 - \gamma_0)(\beta_1 - \gamma_1)\lambda_1 = \det(A_1 - A_0^T).$$

It follows that *two general matrices of order two, together with their transposes, satisfy the condition of the matrix cross-ratio theorem*. The same result is not true for matrices of higher order.

Ex. 2. With the notation of Ex. 1, suppose also that the eigenvalues of A_0 and A_1 are (ψ_0, ψ_0') and (ψ_1, ψ_1') respectively. Show then that the eigenvalues of $A(\lambda)$ are the roots of the equation

$$(\psi - \psi_0)(\psi - \psi_0') + \theta(\psi - \psi_1)(\psi - \psi_1') = 0,$$

where $$\theta = \frac{\beta_0 - \gamma_0}{\beta_1 - \gamma_1} \cdot \frac{\lambda - \lambda^2}{\lambda - \lambda_1}.$$

It follows that, if A_0 and A_1 have no common eigenvalue, then the eigenvalues of $A(\lambda)$ are pairs which correspond in a certain involution. On the other hand if, say, $\psi_1 = \psi_0$, then ψ_0 is also an eigenvalue of $A(\lambda)$.

5. Grassmannian varieties

In this and the following section we shall refer briefly to some of the properties of two types of algebraic variety that will feature largely in the subsequent development. The first of these is the *Grassmannian variety* $G(k, m)$ which is the standard unexceptional model of all the k-dimensional subspaces S_k of a projective space S_m. For detailed accounts of this variety we refer the reader in particular to the memoir [23] of Severi and to Vol. 1, Ch. 7 and Vol. 2, Ch. 14 of Hodge and Pedoe [13].

We shall assume, then, that the reader is familiar with the procedure of representing an S_k of S_m by a point P of S_M, where

$$M = \binom{m+1}{k+1} - 1,$$

the homogeneous coordinates of P being a systematic arrangement of the determinants of order $k+1$ extracted from the $(k+1) \times (m+1)$ matrix of the coordinates of any $k+1$ linearly independent points of the S_k. The locus of P as the S_k varies is the Grassmannian $G(k, m)$ of the S_k's of S_m. The dual Grassmann coordinates of the S_k, obtained in a similar way from the coordinate matrix of any $m - k$ linearly independent primes through S_k, turn out to be essentially only a rearrangement of the coordinates of P, with pairs of these latter interchanged and suitable arrangements as to signs.

$G(k, m)$ is a variety whose dimension d and order ν are given by

$$d = (k+1)(m-k), \quad \nu = \frac{1!\,2!\dots k!}{(m-k)!\,(m-k+1)!\dots m!}\, d!.$$

As examples we may mention

(a) the well-known representation of the lines of S_3 by the points of a (Klein) quadric W_4^2 of S_5,

(b) the mapping of the planes of S_5 on the points of a variety V_9^{42} of S_{19}.

As regards Grassmannians $G(k, m)$ in general, we recall the following facts:

(i) Every Grassmannian $G(k, m)$, by a theorem due to Severi [23], is such that the quadrics containing it have no other common point.

(ii) A pencil of S_k's of S_m (the S_k's that lie in a $[k+1]$ and pass through a $[k-1]$) is represented on $G(k, m)$ by a line; and, more generally:

(iii) There exist on $G(k, m)$ two families of (maximal) linear spaces, namely (a) spaces $[k+1]$ which each represent all the S_k's of S_m that lie in a fixed $[k+1]$, and (b) spaces $[m-k]$ which each represent the S_k's of S_m that pass through a fixed $[k-1]$.

As regards the type of Grassmannian with which we shall be mainly concerned, namely the model $G(n-1, 2n-1)$ of all the (self-dual) spaces $[n-1]$ of S_{2n-1}, we note:

(iv) A Grassmannian $G(n-1, 2n-1)$, being of dimension n^2, is multiply generated by two families of $[n]$'s, each of dimension n^2-1, such that any $[n]$ of the first family represents the $[n-1]$'s of S_{2n-1} that lie in a fixed $[n]$, while any $[n]$ of the second family represents the $[n-1]$'s of S_{2n-1} that pass through a fixed $[n-2]$.

6. Quadratic Veroneseans

Another kind of variety to which we shall frequently make reference in our investigations is that which we call the *quadratic Veronesean transform* of a given variety V, this being simply the projective model of all sections of V by quadrics of the ambient space. More precisely:

If $\xi = (\xi_i)$ $(i = 0, ..., m)$ is the generic point of an irreducible algebraic variety V in S_m, and if $M = \frac{1}{2}(m+1)(m+2)-1$, then the variety V' of S_M with generic point $\eta = (\xi_i \xi_j)$ $(i, j = 0, ..., m)$ is called the *quadratic Veronesean* of V.

The points of V' are in unexceptional birational correspondence with those of V; and if V is non-singular, so is V'. Further, every subvariety V_1' of V' is the quadratic Veronesean of the corresponding subvariety V_1 of V. If V is of dimension d and order ν, then V' (also of dimension d) is of order $\nu' = 2^d \nu$.

If V is a linear space, say S_m itself, then V' is the projective model of the complete system of quadrics of S_m – often referred to as the *Veronesean of quadrics* of S_m, or simply as the *quadric Veronesean* of S_m. It is evidently a variety V'_m of order 2^m whose ambient space is S_M; and the systems of linear subspaces of S_m are transformed into systems of sub-Veroneseans of V'_m. For $m = 1$, 2 and 3, the quadric Veroneseans of S_m are a conic, a Veronese surface $^0F_2^4[5]$ and a quadric Veronesean threefold $V_3'^8[9]$ respectively; and we note, for example, that $V_3'^8[9]$ contains an ∞^3-system (of grade 1) of Veronese surfaces corresponding to the planes of S_3.

The other kind of Veronesean variety with which we shall be concerned is the quadratic Veronesean of a non-singular quadric V_{2n-2}^2 of S_{2n-1}, projective model of all sections of V_{2n-2}^2 by all the other quadrics of S_{2n-1}. This is a variety V'_{2n-2} of order 2^{2n-1} whose ambient space is an S_{M-1}, where $M = n(2n+1) - 1$. It contains two systems of quadric Veronesean $(n-1)$-folds, each of dimension $\frac{1}{2}n(n-1)$, which correspond to the two systems of generating $[n-1]$'s on V_{2n-2}^2. Thus, for example, the quadratic Veronesean of a quadric surface V_2^2 of S_3 is an octavic del Pezzo surface $^1F_2^8$ of the second kind in S_8; and $^1F_2^8$ contains two pencils of conics corresponding to the two systems of generators on V_2^2. Or, again, the quadratic Veronesean of a quadric V_4^2 of S_5 is a 4-fold V'_4 of order 32 in S_{19}; and this contains two ∞^3-systems of Veronese surfaces corresponding to the two systems of generating planes of V_4^2.

CHAPTER 3

CLIFFORD PARALLEL SPACES AND CLIFFORD REGULI

From now onwards, we shall be concerned always with geometry of S_{2n-1} relative to an 'absolute' non-singular quadric primal Ω; and in many of our statements therefore, to avoid tiresome repetition, we shall leave the qualifying phrase 'relative to Ω' to be understood by the reader. In the present Chapter we introduce the general notion of a pair of $[n-1]$'s of S_{2n-1} that are Clifford parallel relative to Ω, and we then discuss in detail the notion of a Clifford regulus as the simplest kind of 1-dimensional system of mutually Clifford parallel $[n-1]$'s of S_{2n-1}.

1. Clifford parallel $[n-1]$'s in S_{2n-1}

Since the definition of Clifford parallel spaces that we propose to adopt will strictly apply only to $[n-1]$'s which meet Ω in non-singular quadrics, we introduce for convenience the following

DEFINITION. An $[n-1]$ of S_{2n-1} will be termed *chordal* (relative to Ω) if it meets Ω in a non-singular quadric (a pair of distinct points if $n = 2$).

Equivalently, we may say that a chordal $[n-1]$ is one which does not touch Ω; or, again, one which does not meet its polar $[n-1]$.

DEFINITION. If Π is a chordal $[n-1]$, then a line which meets Π will be said to be *perpendicular to* Π (relative to Ω) if it also meets the polar space Π' of Π.

Such a line will then also be perpendicular to Π', and all the lines perpendicular to Π will evidently form a linear congruence since a unique transversal to Π and Π' can be drawn through a general point of S_{2n-1}.

Suppose now that Π_1 and Π_2 are a pair of skew chordal $[n-1]$'s and that the polar space Π_2' of Π_2 is also skew to Π_1. Then the perpendiculars to Π_2 from points of Π_1 are the transversal lines of Π_1, Π_2 and Π_2'. By Ch. 2, Prop. 3.2, however, these transversals are the directrix lines of a regulus R with Π_1, Π_2, Π_2' as generators. If it now happens that the polar space Π_1' of Π_1 is also a generator of R, then the directrix lines of R are also perpendicular to Π_1, i.e. Π_1 and Π_2 admit an ∞^{n-1} system of common perpendicular transversals, one through each point of either space. This motivates the

DEFINITION.† Two $(n-1)$-dimensional spaces Π_1 and Π_2 of S_{2n-1} will be said to be *Clifford parallel* (*relative to* Ω) if and only if they and their polar spaces Π_1' and Π_2' are four distinct generators of a regulus R of $[n-1]$'s of S_{2n-1}. They then admit ∞^{n-1} common perpendicular transversals – the directrices of R.

This definition implies, as will be noted, that Π_1, Π_2, Π_1', Π_2' are all chordal and mutually skew; also that if Π_1 is Clifford parallel to Π_2 then it is also Clifford parallel to Π_2'. A space Π cannot be Clifford parallel either to itself‡ or to its polar space Π'; but for convenience in stating certain results we add the subsidiary

DEFINITION. A chordal $[n-1]$ will be said to be *trivially parallel* both to itself and to its polar $[n-1]$.

2. Clifford reguli

We now propose to characterize analytically the aggregate of $[n-1]$'s of S_{2n-1} that are Clifford parallel to a given chordal space Π_0. To this end we choose a coordinate system, as by Ch. 2, §2 we may, so that the equation of Ω is $x^T x + y^T y = 0$ and the reference spaces X and Y are Π_0 and its polar space Π_0'. In these circumstances, since any $[n-1]$ which is Clifford parallel

---

† This definition is an almost word-for-word generalization of Clifford's original definition of parallelism (cf. Ch. 1, §3).

‡ The relation of Clifford parallelism is not only non-reflexive but also, as we shall see shortly, essentially non-transitive.

to X must be skew both to X and Y, it must have an equation of the form $y = Ax$ with $|A| \neq 0$; and the required condition on the matrix A is given by

PROPOSITION 2.1. *When the equation of Ω is $x^T x + y^T y = 0$, the necessary and sufficient condition that the $[n-1]$ given by $y = Ax$ should be Clifford parallel to the reference space X is*

$$A^T A = kI \quad (k \neq 0, -1), \tag{2.1}$$

where k is a scalar.

Proof. Let Π be the $[n-1]$ with equation $y = Ax$. If Π is Clifford parallel to X then (i) Π is skew to X, so that $|A| \neq 0$, (ii) Π is chordal, so that $|A^T A + I| \neq 0$ (cf. Ch. 2, §2(i)), and (iii) the polar space Π' of Π must belong to the regulus R containing X, Y and Π.

By Ch. 2, §2(ii) the equation of Π' is $y = -A^{-1T}x$; and by Ch. 2, §3 the generators of R are given, for different values of the scalar parameter λ, by $y = \lambda Ax$. Hence Π' is a generator of R if and only if

$$-A^{-1T} = \lambda A \tag{2.2}$$

for some value of λ ($\neq \infty$); and the values 0 and 1 of λ are excluded by (i) and (ii) above. A rearrangement of (2.2) now gives (2.1). Also, conversely, if A satisfies (2.1) for some $k \neq 0, -1$, then both A and $A^T A + I$ are non-singular so that Π is skew to X and chordal, and the previous argument then shows that Π is Clifford parallel to X. This completes the proof.

Now consider, in the notation of the above proof, the properties of the regulus R containing X, Y, Π and Π', and represented by the equation

$$y = \lambda Ax. \tag{2.3}$$

If Π_λ is the general $[n-1]$ of R, then its polar space $(\Pi_\lambda)'$ has equation $y = -(\lambda A^T)^{-1}x$, and this can be written by use of (2.1) in the form $y = \lambda' Ax$, where $\lambda' = -(k\lambda)^{-1}$. Thus $(\Pi_\lambda)'$ is the generator $\Pi_{\lambda'}$ of R, where λ, λ' are connected by the involutory relation

$$\lambda\lambda' = -1/k. \tag{2.4}$$

This means, then, that R is *autopolar* for Ω, in the sense that the polar of every generator of R is also a generator of R; and this further implies that every pair of chordal generators of R are either Clifford parallel or (if they are a polar pair) trivially parallel. Finally the two generators of R with parameter values given by $\lambda^2 = -1/k$ are self-polar for Ω, i.e. they lie on Ω; and these two generators separate harmonically in R all the polar pairs of generators of the regulus. These considerations lead to the

DEFINITION. A regulus of $[n-1]$'s of S_{2n-1} which is autopolar for Ω but is not entirely contained in Ω will be called a *Clifford regulus*. The two (distinct) generators of such a regulus that lie on Ω will be called its *terminal generators*.

We make use of the above terminology to recapitulate our results in

THEOREM 2.2. *If two $[n-1]$'s of S_{2n-1} are Clifford parallel, there exists a unique Clifford regulus R which contains them both. The two terminal generators of such a regulus (necessarily distinct and skew) lie on Ω and separate harmonically (within R) all the mutually polar pairs of generators of R. Apart from these two terminal generators, all the other generators of R are chordal and they are all either Clifford parallel or trivially parallel to each other.*

If two spaces Π_1 and Π_2 are Clifford parallel we shall often refer to the Clifford regulus containing them as that which *joins* Π_1 to Π_2. As regards the terminal generators of a Clifford regulus R, we note that these – being necessarily skew – will be generators of the same system or of opposite systems on Ω according as n is even or odd (Ch. 2, § 2, (a), (b), (c)). Thus when n is even Clifford reguli will be of two kinds – α-reguli and β-reguli – while if n is odd there will be no such distinction. Again, for n even, a pair of Clifford parallels may be called α-parallel or β-parallel according as the Clifford regulus to which they belong is an α-regulus or a β-regulus. For n odd no such distinction arises.

Ex. 1. In the case of real elliptic geometry, for which S_{2n-1} is a real projective space and Ω is a virtual quadric, interpret and justify the statements: (i) a pair of Clifford parallel $[n-1]$'s are everywhere equidistant, and (ii) distance as measured between the generators of a Clifford regulus is additive. Property (ii) explains why Wong[27] refers to Clifford reguli as 'additive sets'.

Ex. 2. For the case $n = 2$, show that two lines of S_3 are Clifford parallel if they both meet the same pair of generators of one of the two systems on Ω. The two lines are α-parallel if they meet the same pair of β-generators; and *vice-versa*.

Show also in this case that a Clifford regulus is one regulus of a quadric Q which (a) meets Ω in a skew quadrilateral and (b) is such that Q and Ω are mutually apolar.

If Q satisfies the condition (a) but not (b), show that either regulus on Q is *autoparallel* but not autopolar, i.e. that any two members of it, neither a generator of Ω, are Clifford parallel. [N.B. This illustrates the general point that the autopolar property is, but the autoparallel property is not characteristic of a Clifford regulus.]

Ex. 3. With the notation of the present section, show that every $[n-1]$ Clifford parallel to X is a generator of some quadric of the pencil $x^T x + \lambda y^T y = 0$; also, conversely, that any $[n-1]$-generator of such a quadric that is both chordal and skew to X and Y is Clifford parallel to X. Deduce that the $[n-1]$'s Clifford parallel to a given $[n-1]$ have freedom $\frac{1}{2}(n^2 - n + 2)$.

Show further that the $[n-1]$'s parallel to a given $[n-1]$ that pass through a general point P of S_{2n-1} belong to one or other of two irreducible systems, each of freedom $\frac{1}{2}(n-1)(n-2)$. Discuss the case $n = 2$.

3. Questions of transitivity

The generally non-transitive character of Clifford parallelism follows from

PROPOSITION 3.1. *If, in the notation of the preceding section, $y = A_1 x$ and $y = A_2 x$ are the equations of two spaces Π_1 and Π_2 which are each Clifford parallel to the reference space X, so that*

$$A_1^T A_1 = k_1 I \quad \text{and} \quad A_2^T A_2 = k_2 I \quad (k_1, k_2 \neq 0, -1), \qquad (3.1)$$

then a necessary and sufficient further condition for Π_1 to be Clifford parallel to Π_2 is

$$A_1^T A_2 + A_2^T A_1 = kI \quad (k \neq k_1 + k_2 \text{ or } -(1 + k_1 k_2)). \qquad (3.2)$$

Proof. Suppose first that Π_1 is Clifford parallel to Π_2. Then the polars Π_1' and Π_2' of these spaces have equations $y = -A_1^{-1T}x$ and $y = -A_2^{-1T}x$; and since Π_1, Π_2, Π_1', Π_2' belong to a regulus, we find, by a straightforward application of the matrix cross-ratio theorem (Ch. 2, Th. 4.1) and use of the relations (3.1), that A_1 and A_2 satisfy (3.2).

Now suppose conversely that A_1 and A_2 satisfy (3.2) in addition to (3.1). For the reversion of the algebra, applying the sufficiency part of Th. 4.1 of Ch. 2, it is only necessary to verify that (3.1) and (3.2) together imply that the four spaces Π_1, Π_2, Π_1', Π_2' are all mutually skew. By (3.1) we know that Π_1 and Π_1' are skew, as are also Π_2 and Π_2'.

To verify that Π_1 and Π_2 are skew – and therefore also Π_1' and Π_2' – we observe that

$$(A_1 - A_2)^T (A_1 - A_2) = A_1^T A_1 + A_2^T A_2 - (A_1^T A_2 + A_2^T A_1)$$
$$= (k_1 + k_2 - k)\,I,$$

from which it follows, since $k \neq k_1 + k_2$, that $A_1 - A_2$ is non-singular and hence Π_1 and Π_2 are skew.

Similarly, to see that Π_1 and Π_2' are skew – and hence also Π_2 and Π_1' – we observe that

$$(A_1 + A_2^{-1T})^T (A_1 + A_2^{-1T}) = (A_1^T + k_2^{-1}A_2^T)(A_1 + k_2^{-1}A_2)$$
$$= A_1^T A_1 + k_2^{-2}A_2^T A_2$$
$$+ k_2^{-1}(A_1^T A_2 + A_2^T A_1)$$
$$= k_2^{-1}(k_1 k_2 + 1 + k)\,I,$$

from which, since $k \neq -(1 + k_1 k_2)$, it follows that $A_1 + A_2^{-1T}$ is non-singular and hence Π_1 and Π_2' are skew.

It follows then, since Π_1, Π_2, Π_1', Π_2' are all mutually skew, that the sufficiency section of Th. 4.1 of Ch. 2 can be applied to establish the sufficiency of the conditions (3.1) and (3.2) and this completes the proof.

From the above we derive the following theorem which will be important in the sequel.

THEOREM 3.2. *If Π, Π_1 and Π_2 are any three mutually Clifford parallel $[n-1]$'s of S_{2n-1}, then Π_2 is Clifford parallel to almost all generators of the Clifford regulus joining Π to Π_1.*

Proof. Retaining the previous notation, we take Π to be the reference space X and Π_1, Π_2 to have equations $y = A_1 x$ and $y = A_2 x$ respectively. Then A_1 and A_2, by hypothesis, satisfy conditions (3.1) and (3.2) of Prop. 3.1.

Now let R be the Clifford regulus joining X to Π_1, and let Π_λ be a generator $y = \lambda A_1 x$ of R which is Clifford parallel to X. This excludes X and Y (which are already Clifford parallel to Π_2) and the two terminal generators of R (given by $\lambda^2 = -1/k_1$). By Prop. 3.1 the conditions for Π_λ to be Clifford parallel to Π_2 are those obtained by replacing A_1 by λA_1 in (3.1) and (3.2) with the consequent replacement of k_1 by $\lambda^2 k_1$ and of k by λk. It follows then that Π_λ is Clifford parallel to Π_2 provided only that

(i) $\lambda k \neq \lambda^2 k_1 + k_2$, and

(ii) $\lambda k \neq -(1 + \lambda^2 k_1 k_2)$,

the condition $\lambda^2 \neq -1/k_1$ having already been imposed. There are, therefore, at most six generators of R which are not Clifford parallel to Π_2, namely the two terminal generators of R, the two generators given by (i), which are those not skew to Π_2, and the two generators given by (ii), which are those not skew to the polar space Π'_2 of Π_2. This proves the theorem.

It appears incidentally from our analysis that Π_2 meets at most two generators of the regulus joining Π to Π_1, and moreover it meets them in linear spaces of dimension† at least $\frac{1}{2}n - 1$.

COROLLARY 3.2.1. *If $\Pi_1, \ldots, \Pi_4$ are four mutually Clifford parallel $[n-1]$'s of S_{2n-1}, then the general space of the Clifford regulus joining Π_1 to Π_2 is Clifford parallel to the general space of the Clifford regulus joining Π_3 to Π_4.*

For if Π is the general space of the first regulus, then the Theorem tells us that Π is Clifford parallel to Π_3 and Π_4, and then, further, that Π is Clifford parallel to the general space of the second regulus.

---

† Since $(\lambda A_1 - A_2)^T (\lambda A_1 - A_2) = (\lambda^2 k_1 - \lambda k + k_2) I$, it follows that $\lambda A_1 - A_2$ is singular if and only if $\lambda^2 k_1 - \lambda k + k_2 = 0$; and this is the condition that Π_λ meets Π_2. Moreover, by a well-known inequality concerning the rank of a product of two matrices, it follows that if $\lambda A_1 - A_2$ is singular then its rank is at most $\frac{1}{2}n$; and this means that Π_2 meets Π_λ in a space of dimension at least $\frac{1}{2}n - 1$.

4. Canonical equation of a Clifford regulus

We now propose to find all the Clifford reguli which have a given pair of skew $[n-1]$-generators of Ω as their terminal spaces.

For this purpose we use the second standard form for the equation of Ω, choosing the coordinate system, as by Ch. 2, § 2 we may, so that the two given $[n-1]$'s are the reference spaces X and Y, and the equation of Ω is $x^T y = 0$. If Π is the $[n-1]$ with equation $y = Ax$ (so that Π does not meet Y), we note first that

(i) the polar space Π' of Π has equation $y = -A^T x$, and

(ii) Π is chordal if and only if $A + A^T$ is non-singular.

We now prove

PROPOSITION 4.1. *If A is any symmetric non-singular $n \times n$ matrix, then the equation*

$$y = \lambda Ax \tag{4.1}$$

defines (in the usual way) a Clifford regulus with X and Y as terminal generators. Conversely, every such Clifford regulus is defined by an equation (4.1) in which A is symmetric and non-singular.

Proof. If A is symmetric and non-singular, then $A + A^T = 2A$ is non-singular; whence, by (ii) above, (4.1) defines a regulus R of which every member except X and Y is chordal. Further, since $A = A^T$, the generators of R with parameters $\pm \lambda$ are always polars of one another (cf. (i) above). Hence R is a Clifford regulus.

Conversely, if R is a Clifford regulus with X and Y as terminal spaces, then it is defined by an equation $y = \lambda Ax$, $|A| \neq 0$, and it is such that any two of its generators given by values $\pm \lambda$ of the parameter must be polars of each other. This shows that $A = A^T$, so that A is symmetric; and this completes the proof.

COROLLARY 4.1.1. *The freedom of Clifford reguli with assigned terminal generators is $\frac{1}{2}(n-1)(n+2)$, and that of all Clifford reguli is $\frac{1}{2}(n-1)(3n+2)$.*

We now establish an important canonical form to which the equations of Ω and any given Clifford regulus can be simultaneously reduced.

THEOREM 4.2. *Being given a Clifford regulus R in S_{2n-1}, we can always choose a coordinate system in which (simultaneously) Ω has equation $x^T y = 0$ and R is given by $y = \lambda x$.*

Proof. By taking the terminal generators of R as the reference spaces X and Y, we can certainly reduce the equation of Ω to the form $x^T y = 0$, and then, by Prop. 4.1, R is given by an equation $y = \lambda A x$, where A is symmetric and non-singular. Now there exists a non-singular matrix B such that $A = BB^T$ (since any non-singular symmetric matrix can be so written). In terms of new coordinates (ξ, η) defined by the allowable transformation

$$x = A^{-1}B\xi, \quad y = B\eta,$$

we find that the new equation of Ω is $\xi^T \eta = 0$ and that R is now given by the equation $\eta = \lambda \xi$. This proves the Theorem.

Ex. 1. If Π_0, Π_0' is any (chordal) polar pair of generators of a Clifford regulus R, then they define an involutory collineation $\varpi(\Pi_0, \Pi_0')$ of S_{2n-1} – a biaxial harmonic homology with Π_0 and Π_0' as spaces of united points – which leaves Ω invariant. Prove that $\varpi(\Pi_0, \Pi_0')$ permutes the generators of R among themselves, carrying polar pairs into polar pairs. (Analytically, if Ω and R are taken in the canonical forms given in Th. 4.2, and if Π_0 and Π_0' are given by the parameter values $\lambda = \pm \lambda_0$, then $\varpi(\Pi_0, \Pi_0')$ carries the generator with parameter λ into that with parameter λ_0^2/λ.)

Show also that, if Π_1 and Π_2 are any two distinct chordal generators of R, then there exists a unique polar pair Π_0, Π_0' of generators of R for which the associated homology $\varpi(\Pi_0, \Pi_0')$ carries Π_1 into Π_2.

5. The kernel of a Clifford regulus

Let R be a Clifford regulus and let X, Y denote its terminal generators. Since R, as an algebraic variety, is a V_n^n, it meets Ω (by the Theorem of Bézout) in an $(n-1)$-fold of total order $2n$; but part of this intersection consists of the two terminal generators X, Y, and the remainder is therefore a V_{n-1}^{2n-2}. We propose to discuss the character of this latter variety.

We observe first that any chordal generator Π of R meets Ω in an $(n-2)$-dimensional quadric q; and the directrix line of R through any point of q meets Ω in at least two further points, namely the intersections of the line with X and Y. Thus this line lies entirely on Ω and meets the section of Ω by every chordal generator of R. It follows then that ∞^{n-2} of the directrix lines of R lie on Ω, that these generate the V_{n-1}^{2n-2} referred to above and that they meet every chordal generator of R in the section of Ω by that generator. They will also meet X and Y in $(n-2)$-dimensional quadrics q_X and q_Y which we may call the *terminal quadrics of R*. Recapitulating we have

PROPOSITION 5.1. *If R is a Clifford regulus of S_{2n-1}, then the ∞^{n-2} directrix lines of R that lie on Ω generate a variety V_{n-1}^{2n-2}, the intersection of R with Ω residual to the two terminal generators X, Y of R. These directrix lines meet every chordal generator of R in the section q of that generator by Ω, and they meet X and Y in the terminal quadrics q_X and q_Y of R. All the quadrics q, including q_X and q_Y, lie over any one of themselves in R.*

This leads to the

DEFINITION. *The kernel of a Clifford regulus R of S_{2n-1} is the V_{n-1}^{2n-2} generated by the ∞^{n-2} directrices of R that lie on Ω*, and it will be denoted by the symbol K_R. These same lines will be called the *directrix lines* of K_R, and the non-singular quadrics q in which they meet the generators of R will be called the *quadric-generators* of K_R.

The existence of the terminal quadrics q_X and q_Y of R leads to a further definition which will be of considerable importance in the sequel.

DEFINITION. A couple (Π, q), where Π is an $[n-1]$-generator of Ω and q is a non-singular $(n-2)$-dimensional quadric in Π, will be called an *augmented generator of Ω*. We shall find it more convenient to denote the couple in question by the symbol $\Pi(q)$.

An augmented generator, as will appear, plays the part of a pair of mutually polar $[n-1]$'s that have tended to coincidence in a particular way (within a Clifford regulus).

To obtain equations defining K_R, we may suppose, by Th. 4.2, that Ω and R are given in the canonical forms $x^T y = 0$ and $y = \lambda x$. Then K_R is the locus of points (x, y) that satisfy for some value of λ the equations

$$x^T x = y^T y = 0, \quad y = \lambda x, \tag{5.1}$$

and the terminal quadrics q_X and q_Y (corresponding to $\lambda = 0$ and $\lambda = \infty$) are given by

$$q_X: \quad x^T x = 0, \quad y = 0,$$

$$q_Y: \quad y^T y = 0, \quad x = 0.$$

Now equations (5.1) are readily identified as those of a Segre product of a non-singular $(n-2)$-dimensional quadric with a line – a variety known to be a V_{n-1}^{2n-2} of S_{2n-1} (irreducible for $n > 2$, a pair of skew lines when $n = 2$). This gives

PROPOSITION 5.2. *The kernel K_R of a Clifford regulus R is a Segre product of a non-singular $(n-2)$-dimensional quadric with a line. The terminal generators X and Y of R take on, by the adjunction of the terminal quadrics q_X and q_Y respectively, the character of augmented generators of Ω.*

We now note briefly some further – and easily verifiable – properties of a Clifford regulus R and its kernel K_R.

(i) If d is any directrix line of K_R, and if P is the point where d meets any quadric generator q of K_R, then the polar secundum of d (with respect to Ω) meets the generator of R containing q in the tangent $[n-2]$ to q at P.

(ii) In the notation of (i), as P moves along d the tangent $[n-2]$ to q at P describes a regulus of $[n-2]$'s in the polar secundum of d.

(iii) If q_1, q_2 are any two quadric generators of K_R which lie in generators Π_1, Π_2 of R that are not mutually polar, then the tangent primes to Ω at points of q_1 all touch q_2, i.e. q_2 is the envelope of the $[n-2]$'s in which tangent primes to Ω at points of q_1 meet Π_2.

(iv) In particular, the terminal quadrics of R lie over each other in R in such a way that the tangent prime to Ω at any point

P of either of them touches the other at the point of the latter which lies over P.

This last statement motivates the further

DEFINITION. Two augmented generators $\Pi_1(q_1)$ and $\Pi_2(q_2)$ of Ω will be said to be *compatible* if there exists a Clifford regulus with Π_1, Π_2 as its terminal generators and q_1, q_2 as its terminal quadrics.

In amplification of (iv) then, we leave it to the reader to verify

PROPOSITION 5.3. *Two augmented generators* $\Pi_1(q_1)$ *and* $\Pi_2(q_2)$ *of* Ω *are compatible if and only if* (i) Π_1 *is skew to* Π_2, *and* (ii) *the (non-singular) quadrics* q_1, q_2 *are such that* q_2 *is the envelope of the sections of* Π_2 *by the tangent primes to* Ω *at the points of* q_1.

EX. 1. Prove that two $[n-1]$'s Π_1 and Π_2 of S_{2n-1} are Clifford parallel if and only if (i) they and their polar spaces are all skew to one another, and (ii) they meet Ω in quadrics q_1 and q_2 such that the tangent primes to Ω at points of q_1 all touch q_2. (These conditions then imply that the tangent primes to Ω at points of q_2 all touch q_1.)

EX. 2. Show that a Clifford regulus is uniquely determined by its terminal generators and one of its terminal quadrics; and hence confirm again (cf. Cor. 4.1.1) that the Clifford reguli with given terminal generators have freedom $\frac{1}{2}(n-1)(n+2)$.

EX. 3. If Ω has the equation $x^T y = 0$, then any regulus R with the reference spaces X, Y as two of its generators is given by an equation $y = \lambda A x$ with $|A| \neq 0$; and if R is not entirely contained in Ω it meets Ω residually in a V_{n-1}^{2n-2} which is a Segre product of an $(n-2)$-dimensional quadric with a line. If d is any directrix line of V_{n-1}^{2n-2} and P is the point in which d meets a variable quadric generator q of V_{n-1}^{2n-2}, show that the tangent $[n-2]$ to q at P describes a regulus R_{n-1}^{n-1} lying in a certain secundum through d, but that this is not the polar secundum of d with respect to Ω unless $A = A^T$ so that R is a Clifford regulus.

Show further that if R is not autopolar $(A \neq A^T)$ then it is at least autoparallel if A satisfies the condition

$$A^T A^{-1} + A A^{-1T} = kI$$

for some value of k, a condition which holds always if $n = 2$ but not if $n > 2$.

Ex. 4. If κ is the natural correlation between two skew generators X, Y of Ω (cf. Ch. 2, § 2) and ϖ is a non-singular collineation between X and Y, show that the regulus generated by ϖ is a Clifford regulus R (with X and Y as terminal generators) if and only if the correspondence $\kappa^{-1}\varpi$ in X is a symmetrical polarity. (It will then, of course, be the correspondence of pole and polar with respect to the terminal quadric in X of the Clifford regulus R.) The same condition may be expressed by saying that ϖ must be of the form $\kappa\tau$, where τ is a non-singular symmetric polarity in X.

Ex. 5. If R is the regulus $y = \lambda x$ and A is any non-singular matrix of order n, show that R is a Clifford regulus with respect to the quadric $x^T A y = 0$ if and only if A is symmetric.

Show also that, if C_0 is a fixed non-singular symmetric matrix and S is an arbitrary skew-symmetric matrix, then all the quadrics $x^T(C_0 + S)y = 0$ meet R in the same kernel variety V_{n-1}^{2n-2} (residual to X and Y), but R is a Clifford regulus with respect to only one of them, namely $x^T C_0 y = 0$.

Ex. 6. If n is even and S_1, S_2 are skew-symmetric matrices such that $S_1 - S_2$ is non-singular, show that

$$(y - S_1 x)^T (S_1 - S_2)^{-1}(y - S_2 x) \equiv x^T y$$

and hence, using Prop. 4.1, write down equations for a generic pair of α-parallels.

Similarly, if n is odd and S, T are skew-symmetric matrices such that $I - ST$ is non-singular, show that

$$(x - Ty)^T(I - ST)^{-1}(y - Sx) \equiv x^T y$$

and hence write down equations for a generic pair of Clifford parallels (n odd).

6. Ω-reguli

Besides the Clifford reguli so far considered, we shall also on occasion be concerned with reguli whose generators all lie on

Ω – to be called Ω-*reguli*. Since the generating $[n-1]$'s of either system on Ω are not skew to each other if n is odd, we note that Ω-reguli can only exist in S_{2n-1} when n is even. Further, when n is even, Ω-reguli will be of two kinds, consisting respectively of α-generators and β-generators (of the two systems of generating $[n-1]$'s of Ω). We now assert

PROPOSITION 6.1. *If S is any skew-symmetric non-singular $n \times n$ matrix (so that n is necessarily even) and if the equation of Ω is $x^T y = 0$, then the equation*

$$y = \lambda S x \qquad (6.1)$$

defines an Ω-regulus with the reference spaces X and Y as two of its generators. Conversely every such Ω-regulus is defined by an equation (6.1) in which S is skew-symmetric and non-singular.

Proof. The proof follows at once from the observations that the space with equation $y = Ax$ is skew to X and Y if and only if $|A| \neq 0$, and it lies on Ω if and only if A is skew-symmetric (cf. Ch. 2, §2).

From the above it follows that (i) the Ω-reguli which contain X and Y have freedom $\frac{1}{2}(n-2)(n+1)$, and (ii) the total freedom of Ω-reguli of either system is $\frac{3}{2}(n-2)(n+1)$.

Ex. 1. With the notation of Ex. 4 of the preceding section, show that the regulus generated by a non-singular collineation ϖ between X and Y is an Ω-regulus if and only if the correlation $\kappa^{-1}\varpi$ in X is a (non-singular) skew-polarity. In other words, ϖ must be of the form $\kappa\sigma$, where σ is a non-singular skew polarity in X.

In this case show that a directrix line of the Ω-regulus and its polar secundum meet X in a point and an $[n-2]$ which correspond by σ.

LINEAR SYSTEMS OF CLIFFORD PARALLELS

The general topic of this chapter is the structure and properties of irreducible systems of $[n-1]$'s of S_{2n-1}, of any dimension $r \geqslant 1$, which have the property that their members are in general† Clifford parallel to one another with respect to the absolute quadric Ω.

Among such systems we introduce a special class which we call linear, with the property that every irreducible system of parallels is embedded in a linear system. We then reduce the construction of linear systems of parallels to that of finding solutions to a set of matrix equations—an extension of the Hurwitz–Radon matrix equations. This gives us a canonical form for the equation of any linear system which we study in some detail, especially as regards certain properties which lead us to distinguish a particular kind of linear system, to be called non-singular, as being of maximum interest; and we show how to classify these non-singular linear systems. In a final section we discuss the question as to whether linear systems of parallels with respect to one absolute quadric can also be linear systems of parallels with respect to other absolute quadrics.

1. Clifford parallel and algebraically parallel spaces

Let $\mathscr{G}$ be the irreducible algebraic system of all $[n-1]$'s of S_{2n-1}; and let $\mathscr{G} * \mathscr{G}$ denote the symmetric product of $\mathscr{G}$ by itself, i.e. the irreducible algebraic system of all the *unordered* pairs of spaces of $\mathscr{G}$. We now wish to consider irreducible subsystems of $\mathscr{G} * \mathscr{G}$ such that the generic pair of any one of these subsystems is a pair of Clifford parallel $[n-1]$'s in the strict sense in which we defined this concept in Ch. 3.

† In a system of $[n-1]$'s of dimension $r > 1$, each space will be met by ∞^{r-1} others, so that we cannot expect to find systems of dimension greater than 1 in which every pair of chordal spaces are Clifford (or trivially) parallel. The best that can be hoped for are systems in which *almost all* pairs of spaces are Clifford parallel.

In this connection, however, we must first point out that Clifford parallel pairs in S_{2n-1} are all of one kind when n is odd but of two different kinds if n is even. For, by Th. 2.2 of Ch. 3, the $[n-1]$'s of any Clifford parallel pair are joined by a unique Clifford regulus R; and whereas, if n is odd, R is of the unique type with terminal generators of opposite systems α and β of Ω, we have already noted (Ch. 3, § 2) that, if n is even, R can be either an α-regulus (with terminals α_1, α_2) or a β-regulus (with terminals β_1, β_2). This means then that whereas, for n odd, we can regard all pairs of Clifford parallels as specializations of a single generic pair (Π_1, Π_2), we have to regard all such pairs, when n is even, as specializations of one or other of two generic pairs – a generic pair $(\Pi_1, \Pi_2)^\alpha$ of α-parallels and a generic pair $(\Pi_1, \Pi_2)^\beta$ of β-parallels (cf. Ch. 3, § 5, Ex. 6).

Now suppose, for n odd, that $\mathscr{V}$ is the unique irreducible subsystem of $\mathscr{G} * \mathscr{G}$ defined by the generic parallel pair (Π_1, Π_2). We shall say in this case that two spaces Π_1^* and Π_2^* are *algebraically parallel* if (Π_1^*, Π_2^*) is *any* pair of $\mathscr{V}$, i.e. any specialization of (Π_1, Π_2). Similarly, for n even, if $\mathscr{V}_\alpha$ and $\mathscr{V}_\beta$ are the irreducible subsystems of $\mathscr{G} * \mathscr{G}$ with generic pairs $(\Pi_1, \Pi_2)^\alpha$ and $(\Pi_1, \Pi_2)^\beta$ respectively, we shall say that the spaces of any pair of $\mathscr{V}_\alpha$ are *algebraically α-parallel*, while those of any pair belonging to $\mathscr{V}_\beta$ are *algebraically β-parallel*.

Ex. 1. (*The case $n = 2$.*) The situation for n even can be illustrated for $n = 2$ as follows. If we map the lines of S_3 on the points of a Klein quadric W_4^2 of S_5, then the α-generators and β-generators of Ω are mapped on W_4^2 by the points of two conics k_1 and k_2 respectively – sections of W_4^2 by a skew pair of polar planes π_1 and π_2. We then find that

(i) a pair of α-parallel (resp. β-parallel) lines of S_3 (cf. Ch. 3, § 2, Ex. 2) is represented by a pair of distinct points P_1, P_2 of W_4^2, remote from k_1 and k_2, such that their join $P_1 P_2$ does not lie on W_4^2 but meets π_1 (resp. π_2) and is skew to π_2 (resp. π_1),

(ii) the system $\mathscr{V}_\alpha$ (resp. $\mathscr{V}_\beta$) is represented on W_4^2 by the aggregate of point-pairs of W_4^2 whose joining lines (possibly tangent to W_4^2 or lying on it) meet π_1 (resp. π_2),

(iii) the two types of Clifford regulus in S_3 are represented on W_4^2 by the (proper) conics in which W_4^2 is met by planes which meet one of π_1, π_2 in a line (not touching k_1 or k_2) and the other in a point (not lying on k_1 or k_2).

2. Systems of Clifford parallels

We now introduce the

DEFINITION. An *irreducible system of Clifford parallels* in S_{2n-1} is any irreducible system of $[n-1]$'s, defined over the complex ground field, which is such that a generic pair of its members is a Clifford parallel pair.

If $\mathscr{F}$ is such a system, the above definition implies that its symmetric product $\mathscr{F} * \mathscr{F}$ is contained in the unique irreducible system $\mathscr{V}$ if n is odd, or in one or other of $\mathscr{V}_\alpha$, $\mathscr{V}_\beta$ if n is even; and, further, that the spaces of a generic pair of $\mathscr{F}$ are strictly (not only algebraically) parallel. This again implies that almost all members of $\mathscr{F}$ are chordal; also that two members of $\mathscr{F}$ which do not meet each other or their polar spaces are Clifford parallel. A Clifford regulus is a special type of 1-dimensional irreducible system of Clifford parallels.

An irreducible system $\mathscr{F}$, as above defined, may or may not have the property that it contains the Clifford regulus joining every Clifford parallel pair of its members. This leads to the

DEFINITION. An irreducible system of Clifford parallels is said to be *linear* if it contains the Clifford regulus joining every Clifford parallel pair of its members.

We now prove

THEOREM 2.1. (*Embedding Theorem.*) *Every irreducible system of Clifford parallels is contained in at least one linear system of Clifford parallels.*

Proof. Let $\mathscr{F}$ be the given irreducible system which we suppose to be of dimension $r \geqslant 1$ and let Π_1 and Π_2 be two independent generic spaces of $\mathscr{F}$. Then Π_1 and Π_2, being Clifford parallel, belong to a Clifford regulus. Let Π denote a generic $[n-1]$ of this regulus, and let $\mathscr{F}'$ denote the irreducible system of $[n-1]$'s of S_{2n-1} with Π as generic member. Plainly, then, $\mathscr{F}'$ contains $\mathscr{F}$. Further, by Ch. 3, Cor. 3.2.1, any two independent generic members of $\mathscr{F}'$ are Clifford parallel. If $\mathscr{F}$ was linear to begin with, $\mathscr{F}'$ coincides with $\mathscr{F}$; but, otherwise, $\mathscr{F}'$ strictly contains $\mathscr{F}$, having dimension $r' > r$. Repeating the same argument we find that $\mathscr{F}'$, if it is not linear, is contained in an irreducible system $\mathscr{F}''$ (of Clifford parallels) of dimension $r'' > r'$. Continuation of this process must terminate since the sequence $r, r', r'', \ldots$ is bounded above by the dimension of the complete system of $[n-1]$'s of S_{2n-1}. Hence, after a finite number of steps, we must arrive at a linear system $\mathscr{F}^{(k)}$ which contains the original system $\mathscr{F}$. This proves the Theorem.

We add finally

PROPOSITION 2.2. *A linear system $\mathscr{F}$ of Clifford parallels is autopolar with respect to Ω, i.e. the polar of every space of $\mathscr{F}$ belongs to $\mathscr{F}$.*

The proof may be left to the reader.

3. The canonical equation of a linear system

To simplify the statement of the main theorem of this section we prefix the

DEFINITION. If $r \geqslant 2$, a set $S_1, \ldots, S_{r-1}$ of $r-1$ skew-symmetric matrices of order n will be said to satisfy the *extended Hurwitz–Radon matrix equations for given integers n, r and k $(1 \leqslant k \leqslant r)$* if they are such that

(i) $\qquad S_i^T = -S_i \quad (i = 1, \ldots, r-1),$

(ii) $\qquad S_i^2 = -I \quad \text{if} \quad i \leqslant k-1,$

(iii) $\qquad S_i^2 = 0 \quad \text{if} \quad i > k-1,$

(iv) $\quad S_i S_j + S_j S_i = 0 \quad (i, j = 1, \ldots, r-1, \; i \neq j).$

$$(3.1)$$

The basic result we propose to establish may now be stated as

THEOREM 3.1. *If $\mathscr{F}$ is any r-dimensional linear system $(r \geqslant 1)$ of Clifford parallels in S_{2n-1}, then a coordinate system can be chosen* *in which the absolute quadric Ω has equation $x^T y = 0$ and almost all the $[n-1]$'s of $\mathscr{F}$ (excluding only those which meet the reference space Y) are given, for different sets of values of the parameters $\alpha, \lambda_1, ..., \lambda_{r-1}$ by an equation*

$$y = (\alpha I + \lambda_1 S_1 + ... + \lambda_{r-1} S_{r-1}) x, \qquad (3.2)$$

where $S_1, ..., S_{r-1}$ is a set of $r-1$ linearly independent matrices of order n which satisfy the extended Hurwitz–Radon matrix equations (3.1) for some value of k in the range $1 \leqslant k \leqslant r$.

Conversely, if $S_1, ..., S_{r-1}$ is any such set of matrices, then (3.2) represents almost all the spaces of an r-dimensional linear system of Clifford parallels with respect to the quadric Ω given by $x^T y = 0$.

For the purposes of the proof which follows, we note first that $\mathscr{F}$ must certainly contain pairs of Clifford parallel $[n-1]$'s whose coordinates belong to the ground field; and if we choose any one such pair (Π_1, Π_2), then $\mathscr{F}$ (being linear) must also contain the Clifford regulus R_0 joining Π_1 to Π_2. Further, by Ch. 3, Th. 4.2, we can then choose a coordinate system in which Ω has equation $x^T y = 0$ and R_0 has equation $y = \lambda x$. *In all that follows, therefore, we shall assume that*

   (i) *Ω has equation $x^T y = 0$, and*

   (ii) *$\mathscr{F}$ contains the Clifford regulus R_0 with equation $y = \lambda x$.*

The proof of the first part of Theorem 3.1 proceeds by a sequence of Lemmas involving successive applications of the matrix cross-ratio theorem (Ch. 2, Th. 4.1).

LEMMA A. *The space $y = Ax$ is Clifford parallel to almost all generators of R_0 if and only if $A = \alpha I + T$, where α is a non-zero scalar and T is a skew-symmetric matrix such that $T^2 = tI$, where t is a scalar.*

Proof. This Lemma is valid also when the elements of A belong to an extension of the (complex) ground field. Bearing this in mind, we suppose that $y = Ax$ is Clifford parallel to almost all generators of R_0; and then, if λ is an indeterminate over the field obtained by adjoining the elements of A to the ground field, the space $y = Ax$ will be Clifford parallel to $y = \lambda x$, and hence the four spaces

$$y = Ax, \quad y = -A^T x, \quad y = \lambda x, \quad y = -\lambda x$$

will belong to a regulus. They must therefore be mutually skew; and hence, by the matrix cross-ratio theorem, there must exist a scalar $c(\lambda)$, not equal to 0 or 1, such that†

$$\{\lambda I, A; \; -A^T, \; -\lambda I\} = c(\lambda) I.$$

On reduction, this asserts that A must satisfy the equation

$$\lambda^2 I + \lambda\{1 - 2c(\lambda)\}(A + A^T) + A^T A = 0 \qquad (3.3)$$

involving the indeterminate λ. Replacing λ by another indeterminate μ and subtracting, we see that $A + A^T$ is a scalar multiple of I; for the coefficient of $A + A^T$ after subtraction could only be zero if each of $\lambda\{1 - 2c(\lambda)\}$ and $\mu\{1 - 2c(\mu)\}$ had the same value k, independent of λ and μ, and then (3.3) could not be satisfied. Thus we may write $A + A^T = 2\alpha I$, i.e.

$$A = \alpha I + T,$$

where T is skew-symmetric and α is non-zero (as follows from equation (3.3) since λ is an indeterminate). It now follows from (3.3) that $A^T A$ is also a scalar multiple of I; whence, since

$$A^T A = (\alpha I - T)(\alpha I + T) = \alpha^2 I - T^2,$$

it follows that $T^2 = tI$, where t is a scalar.‡ Having thus proved the necessity of the condition given in the Lemma, we remark that its sufficiency now follows by direct computation; and the Lemma is therefore proved.

For completeness we note that the value of the scalar $c(\lambda)$ is $\{(\lambda + \alpha)^2 - t\}/(4\alpha\lambda)$. Further, we note that the condition of the Lemma does not exclude $T^2 = 0$ (i.e. $t = 0$).

LEMMA B. *If a linear system $\mathscr{F}$ contains both the regulus R_0 and the space $y = (\alpha I + T)x$ as defined in Lemma A, then it contains every space $y = (\delta I + \epsilon T)x$, where δ and ϵ are arbitrary scalars.*

† The order of the elements within the brackets on the left is easiest for computation.

‡ If $t = \alpha^2 \neq 0$, this implying that n is even and $A^T A = 0$, then the polar spaces $y = Ax$ and $y = -A^T x$ meet X in skew spaces $[\frac{1}{2}n - 1]$ (cf. §4, Ex. 2 below).

Proof. It is plainly enough to show that the *generic* space of the type in question belongs to $\mathscr{F}$; and, by the linearity of $\mathscr{F}$, this will follow if we show that the same space belongs to a Clifford regulus joining $y = (\alpha I + T) x$ to a generator of R_0. We therefore interpret δ and ϵ as independent indeterminates over the field obtained by adjoining α and the elements of T to the ground field, and we define μ (to within sign) by the equation

$$\delta^2 - \epsilon\alpha^2 = (1 - \epsilon)(\mu^2 - \epsilon t).$$

By the matrix cross-ratio theorem, we find then that the four spaces

$$y = \mu x, \quad y = -\mu x, \quad y = (\alpha I + T) x, \quad \text{and} \quad y = (\delta I + \epsilon T) x$$

belong to a regulus R; and since the first and third of these are Clifford parallel, while the second is the polar of the first, it follows that R is a Clifford regulus. Hence $\mathscr{F}$, by its linearity, contains R and therefore, *a fortiori*, the space $y = (\delta I + \epsilon T) x$.

LEMMA C. *If a linear system $\mathscr{F}$ contains R_0 and the two spaces $y = (\alpha_1 I + T_1) x$ and $y = (\alpha_2 I + T_2) x$, these latter being (strictly) Clifford parallel, then there exists a scalar t_{12} such that*

$$T_1 T_2 + T_2 T_1 = 2 t_{12} I.$$

Proof. We know, by Lemma A, that $\alpha_1 \neq 0$, $\alpha_2 \neq 0$ and T_1, T_2 are skew-symmetric and such that $T_1^2 = t_1 I$, $T_2^2 = t_2 I$. We express then the condition that the four spaces

$$y = (\pm \alpha_1 I + T_1) x, \quad y = (\pm \alpha_2 I + T_2) x$$

belong to a regulus. By the matrix cross-ratio theorem this reduces to the condition that the matrix

$$[(\alpha_1 + \alpha_2)^2 I - (T_1 - T_2)^2]/(4\alpha_1 \alpha_2)$$

should be a scalar multiple of I; whence, since $T_1^2 = t_1 I$ and $T_2^2 = t_2 I$, it follows that $T_1 T_2 + T_2 T_1 = 2 t_{12} I$, where t_{12} is a scalar.

LEMMA D. *If $y = A_1 x$ and $y = A_2 x$ are two independent generic spaces of a linear system $\mathscr{F}$ which contains R_0, then $\mathscr{F}$ contains also the space $y = (\mu_1 A_1 + \mu_2 A_2) x$ for arbitrary μ_1, μ_2.*

Proof. Since $\mathscr{F}$ is irreducible, it is enough to prove that the space in question belongs to $\mathscr{F}$ when μ_1 and μ_2 are taken to be independent indeterminates over the field obtained by adjoining the elements of A_1 and A_2 to the ground field. By Lemmas A and C we know that

$$A_1 = \alpha_1 I + T_1 \quad \text{and} \quad A_2 = \alpha_2 I + T_2,$$

where $\alpha_1 \alpha_2 \neq 0$ and the skew-symmetric matrices T_1, T_2 satisfy

$$T_1^2 = t_1 I, \quad T_2^2 = t_2 I, \quad T_1 T_2 + T_2 T_1 = 2t_{12} I$$

for suitable t_1, t_2, t_{12}. If we write

$$\mu_1' = \frac{\mu_1}{\mu_1 + \mu_2}, \quad \mu_2' = \frac{\mu_2}{\mu_1 + \mu_2}$$

and take ϕ to be either root of the equation

$$\phi^2 = \mu_1' \alpha_1^2 + \mu_2' \alpha_2^2 - \mu_1' \mu_2' (t_1 + t_2 - 2t_{12}),$$

then $\phi \neq 0$ and we find, with little difficulty, that the space

$$y = (\phi I + \mu_1' T_1 + \mu_2' T_2)\, x$$

is a member of the Clifford regulus determined by the spaces

$$y = (\alpha_1 I + T_1)\, x \quad \text{and} \quad y = (\alpha_2 I + T_2)\, x$$

and so belongs to $\mathscr{F}$. (This amounts to a verification, by the matrix cross-ratio theorem, that the three spaces in question, together with the polar space $y = (-\alpha_1 I + T_1)\, x$ of one of the last two, belong to a regulus.) Hence, by Lemma B, the space

$$y = [\delta I + \epsilon(\mu_1' T_1 + \mu_2' T_2)]\, x$$

belongs to $\mathscr{F}$ for all δ, ϵ. In particular, taking $\delta = \alpha_1 \mu_1 + \alpha_2 \mu_2$ and $\epsilon = \mu_1 + \mu_2$, it follows that $\mathscr{F}$ contains the space

$$y = [(\alpha_1 \mu_1 + \alpha_2 \mu_2) I + \mu_1 T_1 + \mu_2 T_2]\, x = (\mu_1 A_1 + \mu_2 A_2)\, x$$

as required.

Proof of Theorem 3.1. We are now in a position to complete the proof of the first part of the main theorem. We may suppose, as already noted, that the linear system $\mathscr{F}$ of Clifford parallels contains the regulus R_0 with equation $y = \lambda x$. Then almost all members of $\mathscr{F}$ have equations of the form $y = Ax$; and it follows now from Lemma D that the matrices A for which $y = Ax$ belongs to $\mathscr{F}$ form a vector subspace $\mathscr{V}$ of the vector space of all matrices of order n, and the dimension of $\mathscr{V}$ is equal to the dimension r of $\mathscr{F}$. Further, with our choice of coordinate system, it follows that $\mathscr{V}$ contains all scalar multiples of I; and, by Lemma A, every element of $\mathscr{V}$ is of the form $\alpha I + T$, where T is skew-symmetric and such that T^2 is a scalar multiple of I. By Lemmas A and C, therefore, we may choose a basis of linearly independent matrices

$$I, T_1, \ldots, T_{r-1}$$

for $\mathscr{V}$ such that $T_1, \ldots, T_{r-1}$ are skew-symmetric and

$$T_i T_j + T_j T_i = 2t_{ij} I$$

for all i,j (distinct or not), where the t_{ij} are scalars.

To simplify this further, we modify the basis as follows. Let $\mathscr{T}$ denote the matrix (t_{ij}) which is symmetric of order $r-1$ and let the rank of $\mathscr{T}$ be $k-1$. There exists then a non-singular matrix $M = (m_{ij})$, of order $r-1$, such that

$$M \mathscr{T} M^T = -D,$$

where $D = (d_{ij})$ is the diagonal matrix $[1^{k-1}, 0^{r-k}]$. If we now write

$$S_i = \sum_{j=1}^{r-1} m_{ij} T_j \quad (i = 1, \ldots, r-1),$$

then $I, S_1, \ldots, S_{r-1}$ is another basis for $\mathscr{V}$ (with the S_i skew-symmetric); and an easy calculation shows that

$$S_i S_j + S_j S_i = -2d_{ij} I$$

for all i,j. This completes the proof of the first part of Theorem 3.1.

As regards the basis $I, S_1, \ldots, S_{r-1}$ just chosen for $\mathscr{V}$, we note that this is by no means uniquely determined. Thus, if L is any orthogonal matrix of order $k-1$, if M is any non-singular matrix of order $r-k$, and if N is an arbitrary matrix of $k-1$ rows and $r-k$ columns, then the matrix

$$C = (c_{ij}) = \begin{pmatrix} L & N \\ 0 & M \end{pmatrix}$$

has the property $CDC^T = D$, and every C satisfying this equation is of the form indicated. Hence the matrices $S'_1, \ldots, S'_{r-1}$ given by

$$S'_i = \sum_{j=1}^{r-1} c_{ij} S_j \quad (i = 1, \ldots, r-1)$$

satisfy effectively the same equations as $S_1, \ldots, S_{r-1}$.

Turning now to the second (sufficiency) part of Theorem 3.1, we have to prove that, if $S_1, \ldots, S_{r-1}$ satisfy the equations (3.1) for some k ($1 \leqslant k \leqslant r$), then a generic pair of spaces of the system given by (3.2) are Clifford parallel, and that the Clifford regulus joining them is contained in the system. We prove this directly as follows.

Let Π_1 and Π_2 be the generic pair in question, with respective equations

$$y = \left(\alpha I + \sum_{i=1}^{r-1} \lambda_i S_i\right) x,$$

and

$$y = \left(\beta I + \sum_{i=1}^{r-1} \mu_i S_i\right) x,$$

where the $2r$ symbols $\alpha, \beta, \lambda_1, \ldots, \lambda_{r-1}, \mu_1, \ldots, \mu_{r-1}$ denote independent indeterminates. The polar space Π_1' of Π_1 then has equation

$$y = \left(-\alpha I + \sum_{i=1}^{r-1} \lambda_i S_i\right) x.$$

If we make the allowable coordinate transformation

$$\xi = y - \left(-\alpha I + \sum_{i=1}^{r-1} \lambda_i S_i\right) x, \quad \eta = iy - i\left(\alpha I + \sum_{i=1}^{r-1} \lambda_i S_i\right) x,$$

so that Π_1 and Π_1' are now given by $\eta = 0$ and $\xi = 0$ respectively, then the new equation of Ω reduces, by use of equations (3.1), to

$$\xi^T \xi + \eta^T \eta = 0$$

and, if we write

$$\psi = \frac{1}{2\alpha}\left\{(\alpha + \beta)^2 + \sum_{i=1}^{k-1} (\mu_i - \lambda_i)^2\right\},$$

we find with little difficulty that the new equation of Π_2 can be written in the form $\eta = C\xi$, where

$$C = i\psi^{-1}\left\{(\psi - \alpha - \beta) I + \sum_{i=1}^{r-1} (\mu_i - \lambda_i) S_i\right\}.$$

A simple calculation reveals moreover that $C^T C = (2\beta\psi^{-1} - 1)I$. Hence, since $2\beta\psi^{-1} - 1 \neq 0$ or -1, it follows from Prop. 2.1 of Ch. 3 that $\eta = C\xi$ is Clifford parallel to $\eta = 0$ with respect to $\xi^T\xi + \eta^T\eta = 0$, i.e. that Π_2 is Clifford parallel to Π_1 with respect to Ω. Further (cf. Ch. 3, § 2), the equation of the Clifford regulus joining Π_2 to Π_1 is $\eta = \theta C \xi$, where θ is a variable parameter; and, converting back to coordinates (x, y), it is easily verified that the spaces of this regulus all belong to the system defined by equation (3.2). The proof of Theorem 3.1 is thereby complete.†

If we premultiply equation (3.2) by x^T, we obtain

$$x^T y = \alpha x^T x,$$

which shows that, if $\alpha \neq 0$, the space Π given by (3.2) is met in the same quadric by Ω and by the cone $x^T x = 0$. From this, and by using the results in Ch. 3, § 5, we deduce

COROLLARY 3.1.1. *If equation* (3.2) *represents a generator* Π *of* Ω (*i.e. if* $\alpha = 0$), *then* Π *has the same terminal quadric* q (*given by* $x^T x = 0$) *in every Clifford regulus of* $\mathscr{F}$ *of which it is a terminal generator.*

The significance of this result is that Π, by virtue solely of its membership of $\mathscr{F}$, acquires the structure of a unique augmentation $\Pi(q)$ of Π; and we shall express this later (Ch. 6, § 5) by saying that this $\Pi(q)$ *properly belongs* to $\mathscr{F}$.

Finally, in this section, we note that if (for given values of n, r and k) $S_1, ..., S_{r-1}$ is a set of matrices satisfying (3.1), and if Q is any (complex) orthogonal matrix of order n, then

$$Q^T S_1 Q, ..., Q^T S_{r-1} Q$$

is likewise a set of matrices satisfying (3.1) (for the same values of n, r and k) and gives rise, therefore, to a linear system $\mathscr{F}_Q$ of Clifford parallels with equation

$$y = (\alpha I + \lambda_1 Q^T S_1 Q + ... + \lambda_{r-1} Q^T S_{r-1} Q) x. \qquad (3.4)$$

If we then make the allowable coordinate transformation

$$\xi = Qx, \quad \eta = Qy,$$

we see that (3.4) reduces to

$$\eta = (\alpha I + \lambda_1 S_1 + ... + \lambda_{r-1} S_{r-1}) \xi,$$

---

† The algebra of the above proof can be used to show that the Clifford reguli in $\mathscr{F}$ are represented on the affine space of $\alpha, \lambda_1, ..., \lambda_{r-1}$ by conics (of a certain family) having $\alpha = 0$ as a prime of symmetry; but a simpler and more informative derivation of this result comes from the representational theory of Ch. 6 (cf. Ch. 6, § 2, Ex. 3).

while the equation of Ω retains the canonical form $\xi^T \eta = 0$. It follows, therefore, that the system $\mathscr{F}_Q$ is projectively equivalent (by an Ω-preserving collineation) to the system $\mathscr{F}$ of Theorem 3.1. Hence we may record

COROLLARY 3.1.2. *In order to find all projectively distinct linear systems of Clifford parallels (for given n, r and k), it is sufficient to find all the solution-sets of equations (3.1) up to orthogonal equivalence.*

It is possible, however, for two solution-sets to give rise to projectively equivalent linear systems of Clifford parallels, even when they are not orthogonally equivalent.† But, in the interesting cases at least, the simplification afforded by Cor. 3.1.2 is sufficient to reduce the problem of classifying linear systems of Clifford parallels to manageable proportions (cf. § 5 below).

Ex. 1. For $n = 2$, every skew-symmetric matrix is a scalar multiple of

$$S_1 = \begin{pmatrix} 0 & -1 \\ 1 & 0 \end{pmatrix},$$

for which $S_1^2 = -I$. By Th. 3.1, then, we obtain a 2-dimensional linear system of Clifford parallels in S_3 with equation

$$\begin{pmatrix} y_0 \\ y_1 \end{pmatrix} = \begin{pmatrix} \alpha & \lambda_1 \\ -\lambda_1 & \alpha \end{pmatrix} \begin{pmatrix} x_0 \\ x_1 \end{pmatrix}$$

and this is the only 2-dimensional system that can exist (up to projective equivalence). The lines of the system form a linear congruence, being all the transversals of the two skew generators of Ω given by $x_1 = \pm i x_0$, $y_1 = \pm i y_0$ (cf. Ch. 3, § 2, Ex. 2).

For $n = 3$, there exists no non-zero skew-symmetric matrix S such that S^2 is either $-I$ or 0; and there are therefore no linear systems of Clifford parallel planes of S_5 that are more ample than the Clifford regulus.

Ex. 2. If $S_1, ..., S_{r-1}$ are skew-symmetric matrices, anti-commuting by pairs, such that $S_1^2 = ... = S_{r-1}^2 = -I$, show that

---

† If $(S_1, ..., S_{r-1})$ is a solution-set of (3.1), then so also is $(\pm S_1, ..., \pm S_{r-1})$, with arbitrary sign combinations, and the latter is not always orthogonally equivalent to the former. But the two solutions plainly give rise to one and the same linear system of Clifford parallels.

$S_1, \ldots, S_{r-1}$ are necessarily linearly independent. (The same conclusion does not follow from equations (3.1) if $k < r$.)

Ex. 3. If $S_1, \ldots, S_{r-1}$ satisfy equations (3.1), show that each product $S_i S_j$ ($i \neq j$) is a skew-symmetric matrix.

Ex. 4. If $k > 1$ in the equations (3.1), show that n is necessarily even. (N.B. A skew-symmetric matrix of odd order is singular.)

Ex. 5. Further to the proof of the second part of Th. 3.1, establish the following slightly stronger result: *every* space of $\mathscr{F}$ given by the equation (3.2) belongs to some Clifford regulus within $\mathscr{F}$.

4. Non-singular and singular linear systems

If Ω has equation $x^T y = 0$ and if R_0 is the Clifford regulus $y = \theta x$, then Th. 3.1 tells us that any linear ∞^r-system $\mathscr{F}$ of Clifford parallels which contains R_0 has an equation of the form

$$y = (\alpha I + \lambda_1 S_1 + \ldots + \lambda_{r-1} S_{r-1}) x,$$

where $S_1, \ldots, S_{r-1}$ are $r-1$ linearly independent matrices of order n satisfying the extended Hurwitz–Radon equations (3.1) for some value of k in the range $1 \leqslant k \leqslant r$, and this equation covers all the spaces of $\mathscr{F}$ that do not meet the reference space Y. For brevity we shall write

$$S(\lambda) = \lambda_1 S_1 + \ldots + \lambda_{r-1} S_{r-1},$$

so that the equation of $\mathscr{F}$ takes the form

$$y = (\alpha I + S(\lambda)) x. \tag{4.1}$$

We note, by virtue of equations (3.1), that

$$(S(\lambda))^2 = \left(\sum_1^{r-1} \lambda_i S_i \right)^2 = \Lambda^2 I,$$

where $\Lambda^2 = -(\lambda_1^2 + \ldots + \lambda_{k-1}^2)$ if $k > 1$, and $\Lambda = 0$ if $k = 1$. We now make the following observations.

(i) If Π is the space (4.1), then its polar space Π' is given by

$$y = (-\alpha I + S(\lambda)) x.$$

(ii) If $\alpha = 0$, then Π is a generator of Ω.

(iii) If $\alpha \neq 0$, then Π is skew to Π', i.e. Π is chordal.

At this point we must avoid the error of assuming, by virtue of (ii) and (iii) above, that every space of $\mathscr{F}$ must be either chordal or a generator of Ω; for (ii) and (iii) apply only to spaces of $\mathscr{F}$ with equations of the form (4.1), and this excludes spaces of $\mathscr{F}$ which meet the reference space Y. To simplify further discussion of this point we introduce the

DEFINITION. If $\mathscr{F}$ is a linear system of Clifford parallels and if Σ is a terminal generator of a Clifford regulus contained in $\mathscr{F}$, then the spaces of $\mathscr{F}$ which do not meet Σ will be said to form the Σ-affine part of $\mathscr{F}$.

With this terminology we may say that equation (4.1) represents explicitly all the spaces of the Y-affine part of $\mathscr{F}$, and that every space of this Y-affine part is either chordal or a generator of Ω. More generally, allowing for the arbitrary choice of Y in the formulation of Th. 3.1, we have

PROPOSITION 4.1. *If Σ is any terminal generator of a Clifford regulus in the linear system $\mathscr{F}$, then every space in the Σ-affine part of $\mathscr{F}$ is either chordal or a generator of Ω.*

We now introduce the further

DEFINITION. If a space of $\mathscr{F}$ is such that it does not belong to the Σ-affine part of $\mathscr{F}$ for any admissible Σ, then it will be called a *singular space* of $\mathscr{F}$.

It follows then, by Prop. 4.1, that a space of $\mathscr{F}$ that is neither chordal nor a generator of Ω is necessarily a singular space of $\mathscr{F}$. Further, a singular space of $\mathscr{F}$ must meet both terminal generators of every Clifford regulus in $\mathscr{F}$; and, in particular, it cannot belong to any Clifford regulus in $\mathscr{F}$. Conversely, since every space of the Σ-affine part of $\mathscr{F}$ belongs to Clifford reguli contained in $\mathscr{F}$ (cf. §3, Ex. 5) we have

COROLLARY 4.1.1. *A space of $\mathscr{F}$ is singular if and only if it does not belong to any Clifford regulus in $\mathscr{F}$.*

Since singular spaces, by definition, can never be represented explicitly in any canonical equation of $\mathscr{F}$, they are naturally difficult to locate; but examples can be given to show that they do sometimes exist (cf. Ex. 4 below).

We now remark, however, that there are good reasons for distinguishing generally between those linear systems which do not contain singular spaces and those which do. While we are not yet in a position to explain these reasons fully, it may assist the reader at this stage to be given some insight into the general situation. In brief, the key to this is the value of the integer k that occurs in the statement of Th. 3.1. So far we have not proved that k is an invariant† of the linear system $\mathscr{F}$, although this is indeed the case. A direct proof of this, at the present stage of development is straightforward but tedious (involving an extension of the algebra developed in proving the second half of Th. 3.1), and the reader may care to supply his own proof. We prefer, however, to defer this point until Ch. 6, Cor. 2.2.1, where it emerges that k has a natural geometrical meaning relative to $\mathscr{F}$ and so does not depend on any algebraic presentation of the equations of $\mathscr{F}$. Granting then the invariance of k, we introduce the

DEFINITION. If the equation of a linear ∞^r-system $\mathscr{F}$ is given in the canonical form of Th. 3.1, then the number k so arising will be called the *weight* of the system $\mathscr{F}$. If $k = r$, we shall say that $\mathscr{F}$ is *non-singular*; if $k < r$, then $\mathscr{F}$ is *singular*; and in the extreme case (in some ways abnormal) in which $k = 1$, we shall say that $\mathscr{F}$ is *totally singular*.

In terms of this definition, the situation as regards the possible existence of singular spaces of $\mathscr{F}$ may be stated as follows: *a linear system $\mathscr{F}$ possesses singular spaces if and only if it is singular*. The justification of this statement will appear in Ch. 6, §5.

If we take the view, as in general we do, that the most interesting systems of mutually Clifford parallel spaces are those

---

† To show directly that k is an invariant of $\mathscr{F}$, it would be necessary to prove, in connection with the proof of Th. 3.1, that k is independent of the choice of R_0 from among the Clifford reguli contained in $\mathscr{F}$, as also of the choice of the basis I, S_1, ..., S_{r-1} for the vector space $\mathscr{V}$ of matrices associated with $\mathscr{F}$.

which conform reasonably to our intuitive ideas of parallelism, it is found that singular systems can present anomalies. Thus, for example, a singular system may contain pairs of spaces which have common points not 'at infinity', i.e. not on Ω; and we give an example of this below (Ex. 4). Or again, a singular system of Clifford parallel $[n-1]$'s of S_{2n-1} can be such that it is of dimension $r > n$, in which case it must certainly contain pairs of spaces with common points not at infinity; and we give an example of this also (Ex. 5 below). For non-singular systems, however, these anomalies do not arise; and in fact *if two distinct spaces of a non-singular linear system of Clifford parallels intersect, then their intersection lies on Ω*. Here again the proof of this statement must be deferred till we have developed the representational method of Ch. 6; but at this stage we can prove the following weaker result:

PROPOSITION 4.2. *If Π_1 and Π_2 are distinct spaces belonging to the Σ-affine part of a non-singular linear system $\mathscr{F}$ for some admissible Σ, and if Π_1 and Π_2 intersect, then their intersection lies on Ω.*

Proof. Without loss of generality we may take Σ to be the reference space Y, so that the spaces of the Σ-affine part of $\mathscr{F}$ may be taken to be those of the system given by the canonical equation (3.2), in which $S_1, \ldots, S_{r-1}$ satisfy equations (3.1) with $k = r$. We may assume, of course, that neither Π_1 nor Π_2 is a generator of Ω, and hence that Π_1 and Π_2 meet Ω in the quadrics q_1 and q_2 in which they are met by the cone $x^T x = 0$ (cf. the remarks preceding Cor. 3.1.1).

Consider then a point (ξ, η) which is such that $\xi^T \xi \neq 0$ and which lies therefore in neither of q_1, q_2. Any space of the Σ-affine part of $\mathscr{F}$ that passes through (ξ, η) – if any such space exists – will be given by values of $\alpha, \lambda_1, \ldots, \lambda_{r-1}$ that satisfy

$$\eta = (\alpha I + \lambda_1 S_1 + \ldots + \lambda_{r-1} S_{r-1}) \xi. \tag{4.2}$$

If we premultiply (4.2) by $\xi^T, \xi^T S_1, \ldots, \xi^T S_{r-1}$ in turn, we find (using Ex. 3 of §3) that

$$\xi^T \eta = \alpha \xi^T \xi \quad \text{and} \quad \xi^T S_i \eta = -\lambda_i \xi^T \xi; \tag{4.3}$$

and this means, since $\xi^T\xi \neq 0$, that if there exist values of $\alpha, \lambda_1, ..., \lambda_{r-1}$ satisfying (4.2), then these values are uniquely determined by (4.3). Hence any point common to the two (distinct) spaces Π_1, Π_2 must satisfy the equation $x^Tx = 0$, i.e. it must be a point of Ω (common to q_1, q_2). This proves the Proposition.

A further remark arises from the above analysis. If we substitute the values of $\alpha, \lambda_1, ..., \lambda_{r-1}$ given by (4.3) back into (4.2) and replace (ξ, η) by the current coordinates (x, y), we obtain the equation

$$yx^Tx = xx^Ty - S_1xx^TS_1y - ... - S_{r-1}xx^TS_{r-1}y \qquad (4.4)$$

which is satisfied by every point (ξ, η) that lies in a space of the Σ-affine part of $\mathscr{F}$; also, conversely, any point (ξ, η) that satisfies (4.4) and is such that $\xi^T\xi \neq 0$ lies in some space of the Σ-affine part of $\mathscr{F}$. Equation (4.4), equivalent to n ordinary equations, represents a variety in S_{2n-1} which is in general reducible, consisting in part of the variety V generated by all the spaces of $\mathscr{F}$, and residually of one or more components contained in the cone $x^Tx = 0$. The dimension of V is $n-1+r$; and, in particular, if $r = n$, then V must be the whole space S_{2n-1} (equation (4.4) then reducing to an identity). This gives

COROLLARY 4.2.1. *If $n = r = k$ in the canonical equation* (3.2) *of a linear system $\mathscr{F}$, then the spaces of $\mathscr{F}$ form a space-filling system of index* 1.

Ex. 1. Using the identity

$$(\alpha I + S(\lambda))(\alpha I - S(\lambda)) = (\alpha^2 - \Lambda^2)I,$$

where $\Lambda^2 = -(\lambda_1^2 + ... + \lambda_{k-1}^2)$, show that the system $\mathscr{F}$ given by equation (4.1) admits the alternative representation

$$x = (\beta I + \mu_1S_1 + ... + \mu_{r-1}S_{r-1})y,$$

where $\beta, \mu_1, ..., \mu_{r-1}$ are arbitrary parameters. This equation represents explicitly all the spaces of $\mathscr{F}$ that do not meet the reference space X, i.e. the X-affine part of $\mathscr{F}$. (N.B. The condition for (4.1) to meet X is $\alpha^2 - \Lambda^2 = 0$.)

Ex. 2. If $\mathscr{F}$ is a linear system, with generic member Π given by (3.2), and of weight $k > 1$ (so that $S_1^2 = -I$ in (3.1) and n is even), show that Π meets two generators of the Clifford regulus R_0 given by $y = \theta x$ and that it meets each of them in a space $[\frac{1}{2}n - 1]$ which is contained in the kernel of R_0.

[Hint: If $A = (\alpha - \theta) I + S(\lambda)$ then $AA^T = \{(\alpha - \theta)^2 - \Lambda^2\} I$, where $\Lambda^2 \neq 0$. Hence, by well-known rank inequalities, A is of rank $\frac{1}{2}n$ when $\theta = \alpha \pm \Lambda$, and the equations of the two $[\frac{1}{2}n - 1]$'s in question are

$$y = (\alpha \pm \Lambda) x, \quad \{\Lambda I \mp S(\lambda)\} x = 0.]$$

Important deductions from this result will be made in Ch. 5, § 3.

E x. 3. If $\mathscr{F}_1$ and $\mathscr{F}_2$ are two linear systems, of weights k_1 and k_2, such that $\mathscr{F}_1 \subset \mathscr{F}_2$, show that $k_1 \leqslant k_2$.

E x. 4. For $n = 5$, let S_1 denote the skew-symmetric matrix

$$\begin{pmatrix} 0 & 1 & 0 & i & 0 \\ -1 & 0 & i & 0 & 0 \\ 0 & -i & 0 & 1 & 0 \\ -i & 0 & -1 & 0 & 0 \\ 0 & 0 & 0 & 0 & 0 \end{pmatrix}.$$

Verify that $S_1^2 = 0$ and deduce that the ∞^2-system of $[4]$'s of S_9 given by $y = (\alpha I + \lambda_1 S_1) x$ is a (totally singular) linear system $\mathscr{F}$ of Clifford parallels relative to the quadric $x^T y = 0$.

By letting λ_1 tend to ∞ (keeping α finite and non-zero), show that this system $\mathscr{F}$ contains singular spaces, i.e. $[4]$'s which are neither chordal nor generators of Ω.

Show further that those $[4]$'s of $\mathscr{F}$ for which α has a given value all pass through a fixed plane which does not lie on Ω.

E x. 5. Let $n = 2m$, let T be a skew-symmetric matrix of order m whose elements t_{ij} above the diagonal are independent indeterminates, let β be a further indeterminate, and let S be the matrix given by

$$S = \begin{pmatrix} T & -\beta I + iT \\ \beta I + iT & -T \end{pmatrix}.$$

If S' denotes the matrix obtained from S by replacing the t_{ij} and β by t'_{ij} and β', verify that

$$SS' + S'S = -2\beta\beta' I$$

and deduce that $y = (\alpha I + S)x$ is the generic space of a linear system of Clifford parallels (relative to $x^T y = 0$) of freedom $\frac{1}{8}n(n-2) + 2$. [If n is even and $n \geqslant 10$, this implies that there exist (singular) linear systems of Clifford parallel $[n-1]$'s of S_{2n-1} of freedom $r > n$.] The system is of weight $k = 2$ and, for reasons that appear later (Ch. 5, § 3, Ex. 7) we call it a *half-fixed system*.

If we split the column vectors x and y into two halves, writing the equation of the system in the form

$$\begin{pmatrix} y_1 \\ y_2 \end{pmatrix} = \begin{pmatrix} \alpha I + T & -\beta I + iT \\ \beta I + iT & \alpha I - T \end{pmatrix} \begin{pmatrix} x_1 \\ x_2 \end{pmatrix},$$

show that this equation implies the relations:

$$y_1 + iy_2 = (\alpha + i\beta)(x_1 + ix_2),$$

$$x_1^T y_1 + x_2^T y_2 = \alpha(x_1^T x_1 + x_2^T x_2),$$

$$x_1^T y_2 - x_2^T y_1 = \beta(x_1^T x_1 + x_2^T x_2).$$

Deduce that the spaces of the half-fixed system do not fill up S_{2n-1}, and find equations for the locus which they generate.

5. The classification problem

The problem of classifying linear systems of Clifford parallels is equivalent, as we have seen (Cor. 3.1.2) to finding all solution-sets of the extended Hurwitz–Radon matrix equations (3.1) (for given n, r and k) up to orthogonal equivalence; and it would be interesting to know the possible values of n, r and k for which *maximal linear systems* exist (i.e. systems which are not contained in any more ample linear systems). This would appear, however, to be a problem of considerable difficulty and we shall not attempt to solve it in this book.

As we have already indicated, however, there are good reasons why we should be specially interested in *non-singular* linear systems and, for this case at least, the solution of the classification problem can be given. In fact, when $k = r$, equations (3.1) reduce to the well-known and extensively documented Hurwitz–Radon matrix equations and we refer the reader to the Appendix for details. The main result (Appendix, Th. A 1) is as follows:

If n is written in the form $n = u \cdot 2^{4\alpha+\beta}$, with u odd and $0 \leqslant \beta \leqslant 3$, then equations (3.1) – with $k = r$ – admit solutions in matrices of order n if and only if $r \leqslant H(n)$, where $H(n) = 8\alpha + 2^{\beta}$. Moreover there are, up to orthogonal equivalence, only a finite number of solution-sets for given n and r so that, by Cor. 3.1.2, there are only a finite number of projectively distinct non-singular linear systems of Clifford parallels in S_{2n-1}.

The values of $H(n)$ corresponding to low values of n are those given in the table

n	2	3	4	5	6	7	8	9	10	11	12
$H(n)$	2	1	4	1	2	1	8	1	2	1	4

We note also that $H(n) \leqslant n$ for all n. Furthermore, the equality $H(n) = n$ occurs only for $n = 2$, 4 or 8; so that there can (and in fact do) exist *space-filling* non-singular linear systems of Clifford parallels for only these three values of n, i.e. in S_3, S_7 and S_{15}. We shall give much attention in subsequent chapters to explicit geometrical constructions for these space-filling systems (Ch. 5) and to methods of representing them (Ch. 6 and Ch. 7).

Ex. 1. The formula quoted above always gives $H(n) = 1$ when n is odd. This means that, for n odd, there are no non-singular linear systems more ample than the Clifford regulus. Confirm this result by using Ex. 4 of § 3.

Ex. 2. If $n = 8$, $H(n) = 8$ and there accordingly exists a non-singular linear ∞^8-system of Clifford parallel [7]'s in S_{15}. There also exists, however, a non-singular linear ∞^4-system which is *maximal* and hence not embeddable in any ∞^8-system (cf. Ch. 5, § 4).

This illustrates the general point that, if a solution-set of the Hurwitz–Radon equations contains fewer than $H(n)$ matrices, then it is not necessarily possible to augment the solution-set to one containing $H(n)$ matrices.

6. Linear systems $\mathscr{F}$ related to a Clifford kernel

If we are given in S_{2n-1} the absolute quadric Ω and any one Clifford regulus R_0 with respect to Ω, then we know by Th. 3.1 that the set $(\mathscr{F})$ of all linear systems of Clifford parallels with respect to Ω that contain R_0 is determined by the set of all solutions, for all possible values of r and k, of the extended Hurwitz–Radon matrix equations (3.1). We now propose to show that the set $(\mathscr{F})$ depends only on the kernel K_0 of R_0 with respect to Ω; more precisely, that $(\mathscr{F})$ is equally the set of linear systems of Clifford parallels containing R_0 with respect to every non-singular quadric Ω' of a certain net uniquely defined by K_0.

PROPOSITION 6.1. *If Ω has equation $x^T y = 0$, R_0 is the regulus $y = \theta x$, and $\mathscr{F}$ (containing R_0) is the linear system of Clifford parallels with respect to Ω given by equation (3.2), then $\mathscr{F}$ is also a linear system of Clifford parallels with respect to every non-singular quadric of the net*

$$ax^T x + 2bx^T y + cy^T y = 0. \qquad (6.1)$$

Proof. For the purposes of the proof we regard $\alpha, \lambda_1, ..., \lambda_{r-1}$ in (3.2) as independent indeterminates and we write

$$S = \sum_{i=1}^{r-1} \lambda_i S_i \quad \text{and} \quad \Lambda^2 = -(\lambda_1^2 + ... + \lambda_{k-1}^2),$$

so that $S^2 = \Lambda^2 I$.

Suppose then that θ_1, θ_2 are the roots of the equation

$$a + 2b\theta + c\theta^2 = 0$$

(which are distinct if and only if $b^2 \neq ac$, i.e. if and only if (6.1) represents a non-singular quadric) and let (ξ, η) be new co-ordinates in S_{2n-1} related to the coordinates (x, y) by the allowable transformation

$$x = \theta_1 \xi - \eta, \quad y = \theta_2 \xi - \eta.$$

The new equation of the general space $y = \theta x$ of R_0 is then

$$\eta = \phi\xi, \quad \text{where} \quad \phi = \frac{\theta\theta_1 - \theta_2}{\theta - 1},$$

the new equation of $x^T y = 0$ is

$$a\xi^T \xi + 2b\xi^T \eta + c\eta^T \eta = 0, \tag{6.2}$$

and the new equation of $\mathscr{F}$ is

$$\theta_2 \xi - \eta = (\alpha I + S)(\theta_1 \xi - \eta),$$

which can be rearranged in the form

$$\eta = \{(\alpha - 1)I + S\}^{-1}\{(\alpha\theta_1 - \theta_2)I + \theta_1 S\}\xi$$

$$= \{(\alpha - 1)^2 - \Lambda^2)\}^{-1}\{(\alpha - 1)I - S\}\{(\alpha\theta_1 - \theta_2)I + \theta_1 S\}\xi$$

$$= \{(\alpha - 1)^2 - \Lambda^2\}^{-1}\{[(\alpha - 1)(\alpha\theta_1 - \theta_2) - \Lambda^2\theta_1]I + (\theta_2 - \theta_1)S\}\xi;$$

and, since $\alpha, \lambda_1, \ldots, \lambda_{r-1}$ are independent indeterminates, this is equivalent to

$$\eta = (\beta I + \mu_1 S_1 + \ldots + \mu_{r-1} S_{r-1})\xi \tag{6.3}$$

where $\beta, \mu_1, \ldots, \mu_{r-1}$ are new independent indeterminates.

It follows now that $\mathscr{F}$, in the form (6.3), is a linear system of Clifford parallels with respect to the quadric (6.2) for arbitrary values of a, b, c such that $b^2 \neq ac$, and this proves Prop. 6.1.

In amplification of the preceding Proposition, we add the following Corollary, whose proof may be left to the reader.

COROLLARY 6.1.1. *If K_0 is the kernel of the Clifford regulus R_0 (with equation $y = \theta x$) relative to $x^T y = 0$, then every quadric containing K_0 is a member of the system given by*

$$ax^T x + 2bx^T y + cy^T y + x^T S y = 0$$

where S is skew-symmetric; but R_0 is a Clifford regulus only for those non-singular quadrics of this system for which $S = 0$. In other words, when a Clifford regulus and its Clifford kernel are given, the absolute quadric is determined to the extent that it must belong to a certain net.

Finally we prove

PROPOSITION 6.2. *If (Ω) is the net of quadrics uniquely associated, as in Cor. 6.1.1, with a kernel K_0 on R_0, then for $n \geqslant 4$ the base variety of (Ω) is the V_{2n-4}^3 generated by the ∞^{2n-7} solids (not lying in generators of R_0) which meet K_0 in quadric surfaces.*

Proof. We note first that the quadric given by (6.1) is a cone if and only if $b^2 = ac$, and that the equation of any such cone is of the form $(y - \psi x)^T (y - \psi x) = 0$ for some value of ψ. This equation represents the cone projecting K_0 from the generator $y = \psi x$ of R_0. Thus (Ω) is the net of quadrics which contains all the $[n-1]$-cones projecting K_0 from generators of R_0.

Consider then a point P, not on R_0, which lies on all the cones in question. Let Π_1, Π_2 be any pair of generators of R_0 meeting K_0 in quadrics q_1, q_2; and let t be the unique transversal line from P to Π_1, Π_2. Since P lies on the cones $\Pi_1(q_2)$ and $\Pi_2(q_1)$, it follows that t meets Π_1 in a point of q_1 and Π_2 in a point of q_2. Thus P is such that the chords of R_0 which pass through it are all chords of K_0. This means again that the unique quadric solid of R_0 through P is in fact a quadric solid of K_0; and the base locus of (Ω) is therefore generated by quadric solids of K_0.

Now K_0, being the Segre product of an $(n-2)$-dimensional quadric q by a line l, possesses ∞^{2n-7} quadric surfaces (not lying in generators of R_0), each the product of l by a line on q; and the required octavic base locus of the net (Ω) is therefore the V_{2n-4}^8 generated by the solids which meet K_0 in those quadric surfaces. This completes the proof.

Note. The cases $n = 2$ and $n = 3$ are exceptional because K_0 does not possess quadric solids for these values of n. For $n = 2$, the base of the net (Ω) is K_0 itself – a pair of skew lines; and for $n = 3$ the base is K_0 counted twice, K_0 being a rational ruled surface ${}^0R_2^4$.

The case $n = 4$ is special in another way. For in this case K_0, being the Segre product of a quadric surface q by a line l, is also a *Segre product of three lines* (since q itself is a product of two lines); and it therefore possesses two ∞^1-systems of quadric-solids in addition to the systems of generating solids of R_0, i.e. three such systems in all, generating three reguli of which R_0 is one. (We shall return to this later in Ch. 5, § 2, Ex. 3.) Thus for $n = 4$ the base variety V_4^8 of (Ω) breaks up into two V_4^4's – the two reguli other than R_0 associated with K_0 in this case.

For $n > 4$, the lines on a quadric q_{n-2}^2 form a single irreducible system, so that the base V_{2n-4}^8 of (Ω) is irreducible.

Ex. 1. From Prop. 6.1 it appears in particular that, if Ω has equation $x^T x + y^T y = 0$, then any linear system $\mathscr{F}$ of Clifford parallels which contains the Clifford regulus $y = \theta x$ has a canonical equation of the form given in Th. 3.1. This canonical form (for non-singular systems in elliptic space EL_{2n-1}) was originally established by Wong[27] with the equation of Ω in the above form.

Ex. 2. Discuss the character of the net (Ω) which is given for $n = 3$ by the equation

$$a(x_0^2 + x_1^2 + x_2^2) + 2b(x_0 y_0 + x_1 y_1 + x_2 y_2) + c(y_0^2 + y_1^2 + y_2^2) = 0.$$

Chapter 5

GEOMETRICAL CONSTRUCTIONS

In the previous Chapter, our main concern was with the formulation and discussion of an existence problem, namely that of the existence of linear systems – non-singular or singular – of Clifford parallel $[n-1]$'s of S_{2n-1}; and in § 3 of the same Chapter we reduced this problem to its equivalent in terms of matrix equations of which the solutions for the non-singular case were already known. In the present Chapter we turn to more geometrical matters, in particular to the problem of finding explicit geometrical constructions for the linear systems in question. The general construction problem that arises in this connection is one of considerable complexity to which we do not at present know all the answers. It would seem useful, however, to give the details of its solution at least in those cases ($n = 2, 4, 8$) for which space-filling non-singular linear systems exist, and this will in fact be our principal objective in the Chapter. Since the case $n = 2$ has already been explained in detail (Ch. 3, § 2, Ex. 2, Ch. 4, § 1, Ex. 1 and § 3, Ex. 1), we begin with $n = 4$.

1. Clifford parallelism of solids in S_7

We are concerned in this section with Clifford parallelism of solids with respect to a 6-dimensional quadric Ω in S_7 (the case $n = 4$), and our principal objective is to find a simple geometrical construction for the important space-filling type of linear ∞^4-system (of Clifford parallel solids) that exists in this case. Of all the higher dimensional Clifford parallelisms in S_{2n-1}, the case $n = 4$ is easily the most interesting; and both here and in the following Chapters on representation theory we shall treat it in some detail.

Suppose then that X and Y are two skew solids which lie on $\Omega(=\Omega_6)$, and recall (Ch. 2, § 2) that Ω induces a natural correlation κ between X and Y. For convenience of notation we may regard κ as defining a mapping

$$\kappa : \tilde{X} \to \tilde{Y}$$

from the aggregate $\tilde{X}$ of lines of X to the aggregate $\tilde{Y}$ of lines of Y. We begin then by proving

LEMMA A. *Let l and m be two lines, lying in X and Y respectively, which span a solid Π. Then the polar solid Π' of Π (with respect to Ω) is that which spans the lines $l' = \kappa^{-1}(m)$ and $m' = \kappa(l)$.*

Proof. Since l and m' correspond by κ, the solid Σ which they span lies on Ω; and so, likewise, does the solid Σ' spanned by l' and m. Hence since l is the intersection of X with Σ, its polar [5] is the join of X to Σ, and therefore contains l' and m'; and, similarly, the polar [5] of m likewise contains l' and m'. Hence, since Π is spanned by l and m, its polar space Π' contains l' and m', and it is clearly spanned by them. This proves the Lemma.

We now select a non-singular null-polarity σ in X and keep it fixed throughout the following discussion.

Again, for notational convenience, we may regard σ as an (involutory) line–line mapping

$$\sigma \colon \tilde{X} \to \tilde{X}$$

in X. There is then, plainly, an associated null-polarity τ in Y given by

$$\tau = \kappa\sigma\kappa^{-1};$$

and if we put $\varpi = \tau\kappa$, so that also $\varpi = \kappa\sigma$, we see that ϖ (being a product of two correlations) is a collineation from (the lines of) X onto (the lines of) Y. The relations between κ, σ, τ and ϖ (in their capacity as line–line correspondences) may be summed up by saying that the diagram

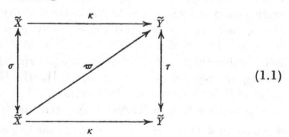

$$(1.1)$$

commutes. We now prove

Lemma B. *If two lines l and m, of X and Y respectively, correspond by the collineation ϖ, then so also do $l' = \sigma(l)$ and $m' = \tau(m)$. Also, the solid spanned by l' and m' is the polar of that spanned by l and m.*

Proof. The first assertion is that, if $m = \varpi(l)$, then

$$\tau(m) = \varpi(\sigma(l)),$$

i.e. that $\tau\varpi(l) = \varpi\sigma(l)$ for every l in X. This follows at once from (1.1).

As regards the second assertion, we know by Lemma A that the polar of the solid (l, m) is that spanned by $\kappa^{-1}(m)$ and $\kappa(l)$; so we have only to show that $\sigma(l) = \kappa^{-1}(m)$ and $\tau(m) = \kappa(l)$, i.e. that $\sigma(l) = \kappa^{-1}\varpi(l)$ and $\tau\varpi(l) = \kappa(l)$. These again follow immediately from (1.1), and the Lemma is proved.

Proposition 1.1. *If l_1 and l_2 are two lines of X such that l_1, l_2, $\sigma(l_1)$ and $\sigma(l_2)$ are four skew lines, and if m_1 and m_2 are their ϖ-images in Y, then the solid Π_1 spanned by l_1 and m_1 is Clifford parallel to the solid Π_2 spanned by l_2 and m_2.*

Proof. By Lemma B, the polar solid Π_1' of Π_1 is spanned by $\sigma(l_1)$ and $\tau(m_1)$, while the polar solid Π_2' of Π_2 is spanned by $\sigma(l_2)$ and $\tau(m_2)$. We have to show that Π_1, Π_2, Π_1', Π_2' belong to a regulus of solids.

Now two lines of S_3 and their images under a null-polarity always belong to a regulus (provided they are mutually skew).[†] So l_1, l_2, $\sigma(l_1)$ and $\sigma(l_2)$ belong to a regulus R_X of lines in X, and m_1, m_2, $\tau(m_1)$ and $\tau(m_2)$ belong to a regulus R_Y in Y; and the collineation ϖ maps R_X into R_Y in such a way that the lines of the former set are mapped respectively into the lines of the latter set. Hence, since X and Y are skew, it follows by Prop. 3.4 of Ch. 2 that the joins (l_1, m_1), (l_2, m_2), $(\sigma(l_1), \tau(m_1))$ and $(\sigma(l_2), \tau(m_2))$ belong to a regulus of solids in S_7, i.e. Π_1, Π_2, Π_1', Π_2' belong to a regulus R. This proves the Proposition.

The regulus R, as will now be apparent, is the Clifford regulus

---

[†] Almost all of what is proved in this section can be generalized to geometry in S_{2n-1}; but the particular assertion just made is the exception, being peculiar to S_3 (cf. Ex. 1 below).

containing Π_1 and Π_2, each generator of R being the join of a line of X to its ϖ-image in Y. Hence, from the definition of a linear system, we obtain at once the following result:

THEOREM 1.2. *Let X and Y be two skew solids on $\Omega = \Omega_6$; let κ be the natural correlation from X to Y induced by Ω; let σ be any non-singular null-polarity in X; and let ϖ denote the collineation $\kappa\sigma$ from X to Y. Then the system $\mathscr{F}$ of ∞^4 solids obtained by joining the ∞^4 lines of X to their respective ϖ-images in Y is a linear system of Clifford parallels.*

The system $\mathscr{F}$ contains in all ∞^6 Clifford reguli. Further, $\mathscr{F}$ contains ∞^3 solids which lie on Ω, namely those solids $(l, \varpi(l))$ for which l is self-polar with respect to σ.

We now observe, further, that the collineation ϖ between X and Y generates a regulus $\bar{R}$ (containing X and Y) whose directrices are the lines joining corresponding points of the two spaces. Also, since ϖ is of the form $\kappa\sigma$, it follows (cf. Ch. 3, § 6, Ex. 1) that $\bar{R}$ is an Ω-regulus. Each solid of $\mathscr{F}$ meets each generator of $\bar{R}$ in a line, i.e. it is a quadric-solid of $\bar{R}$; and since quadric-solids of $\bar{R}$ form an irreducible system of dimension 4, it follows that $\mathscr{F}$ is the system of all quadric-solids of the Ω-regulus $\bar{R}$. Hence:

THEOREM 1.3. *The solids of the system $\mathscr{F}$ defined in Theorem 1.2 are those and only those which meet every generator of a certain Ω-regulus $\bar{R}$ in a line, i.e. they are the quadric-solids of $\bar{R}$.*

We note:

(i) It is implicit in the above that $\mathscr{F}$ determines and is uniquely determined by the Ω-regulus $\bar{R}$. Now (cf. Ch. 3, § 6) the Ω-reguli (for $n = 4$) form two distinct ∞^{15}-families – reguli of α-solids and reguli of β-solids of Ω. It follows, then, that the linear systems of Clifford parallels projectively equivalent to $\mathscr{F}$ form two ∞^{15}-families.

(ii) From Prop. 3.3 of Ch. 2 it follows that *there is a unique solid of the system $\mathscr{F}$ through any point of S_7 that does not lie on $\bar{R}$.* Moreover since, by Th. 1.3, every solid of $\mathscr{F}$ contains ∞^1 mutually skew lines of Ω, it either meets Ω in a non-singular quadric surface or lies entirely on Ω, i.e. *each solid of $\mathscr{F}$ is either chordal or a generator of Ω.*

We remark finally that if Π_1, Π_2, Π_3 are any three generators of $\bar{R}$, then the quadric-solids of $\bar{R}$ are completely identified as all those solids which meet each of Π_1, Π_2, Π_3 in a line. Hence:

The system of all solids which meet three mutually skew generators of Ω in lines is a linear system $\mathscr{F}$ of Clifford parallels, i.e. *it is such a system $\mathscr{F}$ as has been described in Ths. 1.2 and 1.3.*

This gives a simple generation for such systems as $\mathscr{F}$ in terms of incidence properties only.

Ex. 1. Further to the footnote to Prop. 1.1, show that if n is even, if σ is a non-singular skew-polarity in a generator X of Ω, and if $\mathscr{G}$ is an irreducible ∞^r-system of $[\frac{1}{2}n-1]$'s in X which is (a) invariant under σ and (b) such that two general members of it together with their σ-images belong to a regulus, then σ and $\mathscr{G}$ may be used to construct, after the manner of Th. 1.2, an ∞^r-system of Clifford parallel $[n-1]$'s of S_{2n-1}. State a further condition on $\mathscr{G}$ which will ensure that this system of Clifford parallels is linear.

Ex. 2. If Σ is any generator of the regulus $\bar{R}$ mentioned in Th. 1.3, then the spaces of $\mathscr{F}$ are in unexceptional birational correspondence (by incidence) with the lines of Σ. Derive from this a mapping of the spaces of $\mathscr{F}$ on the points of a Klein quadric W_4^2 in S_5, and show that the pairs of polar spaces of $\mathscr{F}$ are represented on W_4^2 by pairs of points whose joins pass through a fixed point not on W_4^2. Obtain from this a mapping of the polar pairs of $\mathscr{F}$ on the points of a space S_4, and discuss the representation in this mapping of the Clifford reguli in $\mathscr{F}$.

2. The case $n = 4$ continued

Arising out of the work of the previous section, two questions remain to be answered.

(a) Is the system $\mathscr{F}$ which we constructed in § 1 a *non-singular* linear system?

(b) Are there any other types of linear system, non-singular or singular, in S_7?

To answer these questions we proceed directly from the equations developed in Ch. 4, the discussion for this case $n = 4$ being simplified by making use of the so-called *dual* matrices and their properties.

We remind the reader that if $S = (s_{ij})$ is a skew-symmetric matrix of order 4, then its dual S^D is defined to be the matrix $T = (t_{ij})$, also skew-symmetric, given by

$$t_{ij} = \frac{1}{2} \sum_{k=1}^{4} \sum_{l=1}^{4} \epsilon_{ijkl} s_{kl},$$

where ϵ_{ijkl} is the alternating symbol on four suffixes. If, in a different notation, we write

$$S = \begin{pmatrix} 0 & -\lambda_1 & -\lambda_2 & -\lambda_3 \\ \lambda_1 & 0 & -\lambda_3' & \lambda_2' \\ \lambda_2 & \lambda_3' & 0 & -\lambda_1' \\ \lambda_3 & -\lambda_2' & \lambda_1' & 0 \end{pmatrix},$$

we find that S^D is obtained from S by interchanging the primed with the unprimed letters. It follows at once that

$$(S^D)^D = S, \tag{2.1}$$

and we verify easily that

$$SS^D = S^D S = -(\lambda_1 \lambda_1' + \lambda_2 \lambda_2' + \lambda_3 \lambda_3') I_4 \tag{2.2}$$

and $$\det S = \det S^D = (\lambda_1 \lambda_1' + \lambda_2 \lambda_2' + \lambda_3 \lambda_3')^2.$$

From (2.2) we deduce that, if S is a non-zero skew-symmetric matrix of order 4 such that S^2 is a multiple of I_4, then S must be†
a multiple of S^D, say $S = kS^D$. Moreover, taking duals of both sides of this equation, we obtain $S^D = kS$, so that $k^2 = 1$ and hence $k = \pm 1$. Hence there are two kinds of skew-symmetric matrix S of order 4 such that S^2 is a multiple of I_4:

(i) matrices S for which $S^D = S$, and
(ii) matrices S for which $S^D = -S$.

---

† Strictly speaking, this deduction only follows from (2.2) if S is non-singular; but it is easy to verify that the conclusion remains valid if S is singular.

Typical matrices of types (i) and (ii) are, respectively,

$$\begin{pmatrix} 0 & -\lambda_1 & -\lambda_2 & -\lambda_3 \\ \lambda_1 & 0 & -\lambda_3 & \lambda_2 \\ \lambda_2 & \lambda_3 & 0 & -\lambda_1 \\ \lambda_3 & -\lambda_2 & \lambda_1 & 0 \end{pmatrix} \text{ and } \begin{pmatrix} 0 & -\lambda_1 & -\lambda_2 & -\lambda_3 \\ \lambda_1 & 0 & \lambda_3 & -\lambda_2 \\ \lambda_2 & -\lambda_3 & 0 & \lambda_1 \\ \lambda_3 & \lambda_2 & -\lambda_1 & 0 \end{pmatrix}. \tag{2.3}$$

Moreover it is easy to check further that

(a) if S_1 and S_2 are both of type (i) or both of type (ii), then $S_1 S_2 + S_2 S_1$ is always a multiple of I_4;

(b) if S_1 and S_2 are of opposite types, then $S_1 S_2 + S_2 S_1$ is never a multiple of I_4 unless $S_1 = 0$ or $S_2 = 0$.

Hence, with reference to Ch. 4, Th. 3.1 and its proof, we deduce the following result:

If Ω has equation $x^T y = 0$ and R_0 is the Clifford regulus $y = \lambda x$, then any linear system of Clifford parallel solids in S_7 that contains R_0 is contained in one or other of the two ∞^4-systems

$$y = (\alpha I + S^{(1)}(\lambda)) x, \quad y = (\alpha I + S^{(2)}(\lambda)) x, \tag{2.4}$$

where $S^{(1)}(\lambda)$ and $S^{(2)}(\lambda)$ are the two matrices given in (2.3) above.

The two systems (2.4) are not, in fact, projectively distinct. Each is transformed into the other if we interchange x_1 with x_2 and y_1 with y_2 (with an interchange also of parameters λ_1 and λ_2), this being equivalent to a self-collineation of S_7 which leaves Ω invariant. It follows that either of the forms (2.4) is a canonical form to which the equations of *any* linear ∞^4-system of Clifford parallel solids in S_7 may be reduced. In particular the equations of the system $\mathscr{F}$ constructed in §1 must be reducible to this form. We therefore have

THEOREM 2.1. *Every linear ∞^4-system of Clifford parallel solids in S_7 is projectively equivalent to the system $\mathscr{F}$ defined by the construction given in Theorems 1.2 and 1.3.*

Moreover, as we have remarked above, any system of Clifford parallel solids in S_7 is contained in a system projectively equivalent to $\mathscr{F}$ and this effectively answers question (b) which we posed at the beginning of this section. As regards question (a), the answer here is in the affirmative, as the reader may easily verify directly from equations (2.4) (cf. also Ex. 1 below). Hence, recapitulating, we have

COROLLARY 2.1.1. *Every linear system, non-singular or singular, of Clifford parallel solids in S_7 is contained in a non-singular linear ∞^4-system which is projectively equivalent to the system $\mathscr{F}$ defined in § 1.*

We might add, as we have already suggested in Ch. 4 (§§ 4, 5), that simple and concise results of the above kind are not to be expected in general for linear systems of Clifford parallels in S_{2n-1}.

Ex. 1. Using the notation defined in the Appendix, § 3, show that the matrices $S^{(1)}(\lambda)$ and $S^{(2)}(\lambda)$ in (2.4) can be expressed in the forms

$$S^{(1)}(\lambda) = \lambda_1 I \times J + \lambda_2 J \times K + \lambda_3 J \times L,$$
$$S^{(2)}(\lambda) = \lambda_1 K \times J + \lambda_2 J \times I + \lambda_3 L \times J,$$

and hence verify directly that the two systems given by (2.4) are non-singular.

Ex. 2. By the result given in Th. 1.3, there must exist Ω-reguli associated as in this Theorem with the two systems given by (2.4). Verify that these Ω-reguli are given, with $\epsilon = 1$ and $\epsilon = -1$ respectively, by the equations

$$x_0 + ix_1 = \lambda(x_2 + \epsilon ix_3),$$
$$x_0 - ix_1 = -\lambda^{-1}(x_2 - \epsilon ix_3),$$
$$y_0 + iy_1 = \lambda(y_2 + \epsilon iy_3),$$
$$y_0 - iy_1 = -\lambda^{-1}(y_2 - \epsilon iy_3).$$

Ex. 3. A Clifford regulus R in S_7 is contained in just two (non-singular) linear ∞^4-systems of Clifford parallel solids. The Ω-reguli with which these are associated (as in Th. 1.3) consist of solids of one and the same system on Ω. In fact, as we observed in Ch. 4, § 6, a notable feature of the case $n = 4$ is that the kernel of a Clifford regulus R, being the Segre product of a quadric surface with a line, is therefore the product of three lines; so that it may be regarded in *three separate ways* as the product of a quadric surface by a line. It lies, therefore, on three different reguli of solids of which one is the Clifford regulus R from which we started. The other two are the Ω-reguli associated with the two linear ∞^4-systems of Clifford parallels that contain R.

3. The Bridging Theorem

Having now discussed the case $n = 4$ in some detail, our next objective is to find geometrical constructions for systems of Clifford parallel [7]'s in S_{15} (the case $n = 8$), and most particularly for the third and last type of space-filling non-singular linear system – an ∞^8-system of [7]'s in S_{15}. In preparation for the subsequent attack on this problem, however, we first exhibit in the present section a general line of procedure for constructing linear systems of Clifford parallel $[n-1]$'s for any even value of n. This procedure as it happens, apart from its subsequent application to S_{15}, will also give us (cf. Ex. 5 below) a new construction for the space-filling linear systems of solids in S_7 – a construction that is no less simple, though quite different from that which we established in § 1.

Suppose then that n is even and that R is a Clifford regulus in S_{2n-1} and let K_R be the kernel of R. Further, let $\mathscr{F}$ be a linear system of Clifford parallels containing R. Then, as we have already noted (Ch. 4, § 4, Ex. 2), if $\mathscr{F}$ has weight $k > 1$, the generic space Π_0 of $\mathscr{F}$ meets R in two linear spaces $\bar{\Sigma}_0$ and $\bar{\Sigma}'_0$, each of dimension $\frac{1}{2}n - 1$, which lie on K_R. *We now select for purposes of reference a fixed quadric generator q of K_R.* In the language of Ch. 3, Prop. 5.1, the two spaces $\bar{\Sigma}_0$ and $\bar{\Sigma}'_0$ then lie *over* a pair of generators Σ_0 and Σ'_0 of q, and these can be shown to be skew to one another (cf. Th. 3.1 below). In this situation we shall say that the space Π_0 *bridges* the pair of generators (Σ_0, Σ'_0) of q; it is, in fact, one of ∞^2 $[n-1]$'s of S_{2n-1} which bridge the pair (Σ_0, Σ'_0). The result given in Ch. 4, § 4, Ex. 2 may now be expressed by saying that the generic space Π_0 of a linear system $\mathscr{F}$, of weight $k > 1$, bridges† a non-intersecting pair of generators

---

† We emphasize that the role of q in this statement is only that of a selected base of reference, and any other quadric generator of K_R would serve the same purpose equally well. Further, the bridging phenomenon takes place relative to *any* quadric generator of *any* Clifford kernel within the system $\mathscr{F}$.

(Σ_0, Σ_0') of q. We propose now to improve this statement as follows.

THEOREM 3.1. *Let n be even; let $\mathscr{F}$ be a linear system of Clifford parallels with respect to Ω, containing the Clifford regulus R; and let q be any quadric generator of the kernel K_R of R. Then*

(i) *if the weight k of $\mathscr{F}$ is greater than 1, a generic space Π_0 of $\mathscr{F}$ bridges a non-intersecting pair (Σ_0, Σ_0') of generators of q;*

(ii) *if a second generic space Π_1 of $\mathscr{F}$ (independent of Π_0) bridges the pair (Σ_1, Σ_1'), then a sufficient condition for $\Sigma_0, \Sigma_0', \Sigma_1, \Sigma_1'$ to be four mutually skew spaces is $k > 2$ (a condition which is also necessary when $\mathscr{F}$ is maximal†); and*

(iii) *when this condition is satisfied, $\Sigma_0, ..., \Sigma_1'$ belong to a regulus of generators on q.*

Proof. By Th. 3.1 of Ch. 4 we may take the equation of Ω to be $x^T y = 0$, that of R to be $y = \lambda x$, and that of a generic space Π_0 of $\mathscr{F}$ to be

$$y = (\alpha I + \lambda_1 S_1 + ... + \lambda_{r-1} S_{r-1}) x = \{\alpha I + S(\lambda)\} x, \qquad (3.1)$$

where $\alpha, \lambda_1, ..., \lambda_{r-1}$ are independent indeterminates and $S_1, ..., S_{r-1}$ satisfy the extended Hurwitz–Radon matrix equations (3.1) of Ch. 4. We may then choose q to be the terminal quadric of R given by

$$y = 0, \quad x^T x = 0.$$

We now find easily (cf. Ch. 4, §4, Ex. 2) that Π_0 bridges the pair of generators (Σ_0, Σ_0') of q with equations

$$\Sigma_0: \quad \{\Lambda I + S(\lambda)\} x = 0, \qquad (3.2)$$

$$\Sigma_0': \quad \{\Lambda I - S(\lambda)\} x = 0, \qquad (3.2)'$$

where $\Lambda^2 = -(\lambda_1^2 + ... + \lambda_{k-1}^2)$.‡ Since any common solution of (3.2) and (3.2)' satisfies $\Lambda x = 0$, it follows that $\Lambda = 0$ is a necessary (and sufficient) condition for Σ_0 and Σ_0' to meet. But since $\alpha, \lambda_1, ..., \lambda_{r-1}$ are independent indeterminates, we see that

---

† The proof of this assertion of necessity will be deferred until later. See Ex. 4 at the end of this section and Ch. 6, §2, Ex. 1.

‡ Each of (3.2) and (3.2)' represents n linear equations in $x_0, ..., x_{n-1}$; but these n equations reduce to $\frac{1}{2}n$ equations because each of the matrices $\Lambda I \pm S(\lambda)$ is of rank $\frac{1}{2}n$.

$\Lambda^2 = -(\lambda_1^2 + \ldots + \lambda_{k-1}^2)$ is zero if $k = 1$ but non-zero if $k > 1$. This proves (i) of the Theorem.

Now suppose that Π_1 has equation

$$y = (\alpha'I + \lambda_1'S_1 + \ldots + \lambda_{r-1}'S_{r-1})\,x = \{\alpha'I + S(\lambda')\}x,$$

where $\alpha', \lambda_1', \ldots, \lambda_{r-1}'$ are further independent indeterminates, independent of $\alpha, \lambda_1, \ldots, \lambda_{r-1}$. Then Π_1 will bridge the pair (Σ_1, Σ_1') of generators of q given by

$$\Sigma_1: \quad \{\Lambda'I + S(\lambda')\}x = 0, \tag{3.3}$$

$$\Sigma_1': \quad \{\Lambda'I - S(\lambda')\}x = 0, \tag{3.3'}$$

where $\Lambda'^2 = -(\lambda_1'^2 + \ldots + \lambda_{k-1}'^2)$. If Σ_0 were to meet Σ_1 (resp. Σ_1'), then (3.2) and (3.3) (resp. (3.3)') would have a common solution for x; and then, if

$$T = \{\Lambda I + S(\lambda)\} + \{\Lambda'I \pm S(\lambda')\},$$

any such common solution would also satisfy $Tx = 0$, so that T would be singular. But if we write

$$\phi = \lambda_1\lambda_1' + \ldots + \lambda_{k-1}\lambda_{k-1}',$$

we verify easily (using the relations (3.1) of Ch. 4, §3) that

$$TT^T = 2(\Lambda\Lambda' \pm \phi)\,I;$$

so that T is singular if and only if $\Lambda\Lambda' \pm \phi = 0$. Since this equation is not satisfied (with either choice of sign) when $k > 2$ and the λ_i, λ_j' are independent indeterminates, it follows that Σ_0 does not meet Σ_1 or Σ_1' when $k > 2$; and a similar argument applies to Σ_0' and Σ_1 or Σ_1'. This proves (ii) of the Theorem.

Finally, to show that, when $k > 2$, the four spaces $\Sigma_0, \Sigma_0', \Sigma_1, \Sigma_1'$ belong to a regulus it is only necessary to prove that every line which meets Σ_0, Σ_0' and Σ_1 also meets Σ_1', i.e. if $x^{(1)}$ satisfies (3.2), if $x^{(2)}$ satisfies (3.2)' and if $x^{(1)} + \alpha x^{(2)}$ satisfies (3.3), then there exists a scalar β such that $x^{(1)} + \beta x^{(2)}$ satisfies (3.3)'. The verification of this, amounting to an exercise in the properties of the extended Hurwitz–Radon matrix equations, may be left to the reader (cf. Ex. 1 below): and we note only that the appropriate value of β is given by

$$\beta = \frac{\phi - \Lambda\Lambda'}{\phi + \Lambda\Lambda'}\alpha.$$

This proves (iii), and the proof of Th. 3.1 is therefore complete.

We now investigate the converse situation and prove

THEOREM 3.2. (*The Bridging Theorem.*) *Let* n *be a multiple*† *of 4 and let* q *be any quadric-generator of the kernel* K_R *of a Clifford regulus* R *with respect to* Ω. *Then any space* Π_0 *which bridges a skew pair* (Σ_0, Σ_0') *of generators of* q *is Clifford parallel to almost all spaces of* R. *Further, if a second space* Π_1 *bridges a skew pair* (Σ_1, Σ_1') *of generators of* q, *and if* Π_1 *is skew both to* Π_0 *and to its polar space* Π_0', *then* Π_1 *is Clifford parallel to* Π_0 *if* $\Sigma_0, \Sigma_0', \Sigma_1, \Sigma_1'$ *belong to a regulus.*

(It will be noted that we have not precluded the possibility that Σ_1 may coincide with Σ_0 or Σ_0', nor even, more specially, that the pair (Σ_1, Σ_1') may coincide with the pair (Σ_0, Σ_0').)

Proof. We shall prove the second of the two assertions in the theorem, observing that this includes the first as a special case because any generator of R bridges every pair of skew generators of q.

Suppose then that the spaces $\Sigma_0, \Sigma_0', \Sigma_1, \Sigma_1'$ to which the Theorem refers belong to a regulus R_0. *We may assume then, without any loss of generality, that* R_0 *lies on* q. For if three at least of the spaces are distinct, then R_0 must certainly lie on q (since all its directrix lines must do so); and if, on the other hand, the pair (Σ_1, Σ_1') coincides with the pair (Σ_0, Σ_0'), then we can take R_0 to be any one of the reguli on q containing Σ_0 and Σ_0' (of which there are infinitely many since n is a multiple of 4).

We now choose a coordinate system (in S_{2n-1}) which, it may be noted, is not of either of the two standard kinds we have so far employed. First, since n is even, we may split up the $2n$ homogeneous coordinates into four column vectors ξ, ξ', η, η', each of length $\frac{1}{2}n$. We then choose two distinct generators, π and π' say, of the q-regulus R_0, and we arrange the coordinate system, as we evidently may, so that

(i)   the ambient space of q is given by $\eta = \eta' = 0$;

(ii) the equation of q (in this ambient space) is $\xi^T \xi' = 0$; and

(iii) the equations of π and π' (in the same space) are $\xi = 0$ and $\xi' = 0$ respectively.

† For the case when n is an odd multiple of 2, see Theorem 3.2′ and the remarks preceding it.

Next, by an allowable coordinate transformation which does not affect (i), (ii) and (iii), we arrange further that

(iv) the generators of the Clifford regulus R are given, for variable μ, by the equations $\eta = \mu\xi$, $\eta' = \mu\xi'$.

We now recall (cf. Ch. 4, Cor. 6.1.1) that when a Clifford regulus and its kernel are given, the absolute quadric Ω is determined to the extent that it must belong to a certain net which, in the present coordinate system, is easily found to have the equation

$$a\xi^T\xi' + b(\xi^T\eta' + \eta^T\xi') + c\eta^T\eta' = 0; \qquad (3.4)$$

and we may suppose, therefore, that (3.4) is the equation of Ω for certain fixed values of a, b and c such that $ac \neq b^2$ (the condition of non-singularity). With this equation for Ω, we find, then, that the polar of the generator of R with parameter μ is that with parameter $-(a+b\mu)/(b+c\mu)$; and the parameters of the terminal generators of R are the roots of the equation $c\mu^2 + 2b\mu + a = 0$.

Finally, we make the allowable transformation to new co-ordinates (x, x', y, y') given by

$$\left.\begin{array}{ll} x = \alpha\xi + \beta\eta, & x' = \alpha\xi' + \beta\eta' \\ y = \gamma\xi + \delta\eta, & y' = \gamma\xi' + \delta\eta' \end{array}\right\} \quad (\alpha\delta \neq \beta\gamma),$$

where α, β, γ, δ are chosen to satisfy

$$\begin{pmatrix} \alpha & \gamma \\ \beta & \delta \end{pmatrix} \begin{pmatrix} \alpha & \beta \\ \gamma & \delta \end{pmatrix} = \begin{pmatrix} a & b \\ b & c \end{pmatrix},$$

this choice being always possible because $ac \neq b^2$. The resulting new equation of Ω is then

$$x^Tx' + y^Ty' = 0, \qquad (3.5)$$

and the new equations of R are

$$y = \lambda x, \quad y' = \lambda x',$$

the parameter λ being related to the parameter μ of (iv) above by the equation $\lambda = (\delta\mu + \gamma)/(\beta\mu + \alpha)$. Further, the generator of R which contains q is now given by the value $\gamma/\alpha = \theta$, say, of λ; the equation of q in this generator is

$$x^Tx' = 0;$$

and the equations of π and π' are $x = 0$ and $x' = 0$ respectively. The generators of the regulus R_0 on q will now be given (cf. Ch. 3, § 6) by an equation of the form

$$x' = \psi S x, \tag{3.6}$$

where S is a fixed non-singular skew-symmetric matrix of order $\frac{1}{2}n$ and ψ is a parameter. We may suppose, therefore, that the equations of Σ_0, Σ_0', Σ_1, Σ_1' (inside the space $y = \theta x$, $y' = \theta x'$) are given by substituting ψ_0, ψ_0', ψ_1, ψ_1' for ψ in (3.6).

The lengthy process of choosing a convenient coordinate system now pays its dividend, for the rest of the proof is quite straightforward. The space lying over Σ_0 in the generator of R given by $\lambda = \lambda_0$ plainly has equations

$$y = \lambda_0 x, \quad y' = \lambda_0 x', \quad x' = \psi_0 S x,$$

while that over Σ_0' in the generator given by $\lambda = \lambda_0'$ has equations

$$y = \lambda_0' x, \quad y' = \lambda_0' x', \quad x' = \psi_0' S x.$$

The bridging join Π_0 of these two spaces has equation

$$\begin{pmatrix} x' \\ y' \end{pmatrix} = \begin{pmatrix} \alpha_0 S & \beta_0 S \\ \gamma_0 S & \delta_0 S \end{pmatrix} \begin{pmatrix} x \\ y \end{pmatrix} \tag{3.7}$$

where
$$\left. \begin{aligned} \alpha_0 &= \frac{\lambda_0 \psi_0' - \lambda_0' \psi_0}{\lambda_0 - \lambda_0'}, & \beta_0 &= \frac{\psi_0 - \psi_0'}{\lambda_0 - \lambda_0'}, \\ \gamma_0 &= -\lambda_0 \lambda_0' \frac{\psi_0 - \psi_0'}{\lambda_0 - \lambda_0'}, & \delta_0 &= \frac{\lambda_0 \psi_0 - \lambda_0' \psi_0'}{\lambda_0 - \lambda_0'}. \end{aligned} \right\} \tag{3.8}$$

The polar Π_0' of Π_0 with respect to Ω (as given by (3.5)) then has equation
$$\begin{pmatrix} x' \\ y' \end{pmatrix} = \begin{pmatrix} \alpha_0 S & \gamma_0 S \\ \beta_0 S & \delta_0 S \end{pmatrix} \begin{pmatrix} x \\ y \end{pmatrix}. \tag{3.9}$$

Similarly, the equation of any space Π_1 bridging Σ_1 and Σ_1' is of the form
$$\begin{pmatrix} x' \\ y' \end{pmatrix} = \begin{pmatrix} \alpha_1 S & \beta_1 S \\ \gamma_1 S & \delta_1 S \end{pmatrix} \begin{pmatrix} x \\ y \end{pmatrix} \tag{3.10}$$

and its polar space Π_1' is given by
$$\begin{pmatrix} x' \\ y' \end{pmatrix} = \begin{pmatrix} \alpha_1 S & \gamma_1 S \\ \beta_1 S & \delta_1 S \end{pmatrix} \begin{pmatrix} x \\ y \end{pmatrix}. \tag{3.11}$$

Finally, since by hypothesis Π_0, Π_0', Π_1, Π_1' are mutually skew, a simple application of the matrix cross-ratio theorem (cf. also Ex. 2 below) confirms that the four spaces given by (3.7), (3.9), (3.10) and (3.11) belong to a regulus. Hence Π_1 is Clifford parallel to Π_0, and the proof of the Bridging Theorem is complete.

It will be noted that the parameter θ which determines the generator R containing q does not appear in any of the above equations; and this reflects the fact, already noted, that any other quadric-generator of K_R can be substituted for q. We may also draw attention to the relatively minor role of Ω in the Bridging Theorem. We see, in fact, that the Clifford parallelism is essentially determined by the Clifford kernel K_R of R, and that Ω plays thereafter no essential role. This agrees with what we have already said in Ch. 4, § 6.

More importantly, we note that if λ_0 and λ_0' in (3.8) are replaced by $-1/\lambda_0'$ and $-1/\lambda_0$ respectively, then α_0 and δ_0 remain unaltered while β_0 and γ_0 are interchanged, i.e. the substitution converts (3.7) into (3.9), these being the equations representing Π_0 and Π_0' respectively. Since the parameters λ, λ' of any polar pair of generators of R with respect to Ω are such that $\lambda\lambda' = -1$, we have the following result:

COROLLARY 3.2.1. *A space Π_0 and its polar space Π_0' bridge the same pair (Σ_0, Σ_0') of generators of q. More particularly, if Π_0 is the join of the spaces over Σ_0 and Σ_0' that lie in generators T_1 and T_2 of R, and if T_1' and T_2' are the polar generators of T_1 and T_2 in R, then Π_0' is the join of the spaces over Σ_0 and Σ_0' that lie in T_2' and T_1' respectively.*

We next note the easily verifiable algebraic result:

COROLLARY 3.2.2. *If a 2×2 matrix*

$$A_0 = \begin{pmatrix} \alpha_0 & \beta_0 \\ \gamma_0 & \delta_0 \end{pmatrix}$$

can be parametrized in the form (3.8), then ψ_0 and ψ_0' are the eigenvalues of A_0.

The spaces Π_0 and Π_1, given by (3.7) and (3.10), being Clifford parallel, must belong to a Clifford regulus R; and it is natural to ask which pairs of generators of q are bridged by the other spaces of R. To answer this we observe that the generic space of R has equation

$$\begin{pmatrix} x' \\ y' \end{pmatrix} = \begin{pmatrix} \alpha S & \beta S \\ \gamma S & \delta S \end{pmatrix} \begin{pmatrix} x \\ y \end{pmatrix}, \qquad (3.12)$$

where

$$\begin{pmatrix} \alpha & \beta \\ \gamma & \delta \end{pmatrix} = A(\lambda)$$

is the 2×2 matrix, depending on the single parameter λ, defined in Ch. 2, §4, Ex. 1 (Π_0 and Π_1 being given respectively by $\lambda = 0$ and $\lambda = \infty$). From the algebra of the Bridging Theorem it follows that the space (3.12) bridges two generators of the regulus R_0 given by $x' = \psi Sx$, the two appropriate values of ψ being, by Cor. 3.2.2 above, the eigenvalues of $A(\lambda)$. As λ varies, however (cf. Ch. 2, §4, Ex. 2), the eigenvalues of $A(\lambda)$ are the roots of the equation

$$(\psi - \psi_0)(\psi - \psi_0') + \theta(\psi - \psi_1)(\psi - \psi_1') = 0,$$

where θ is a certain function of λ. If ψ_0, ψ_0', ψ_1 and ψ_1' are all distinct, the pairs of roots of this equation, as λ (and hence θ) varies, are pairs of mates in the unique involution determined by the pairs (ψ_0, ψ_0') and (ψ_1, ψ_1'). This gives

COROLLARY 3.2.3. *If Π_0 and Π_1 are two Clifford parallel $[n-1]$'s bridging respectively the pairs of generators (Σ_0, Σ_0') and (Σ_1, Σ_1') of a regulus R_0 on q (the spaces $\Sigma_0, ..., \Sigma_1'$ being distinct), then any other space Π_2 of the Clifford regulus joining Π_0 to Π_1 bridges a pair (Σ_2, Σ_2') of generators of R_0 belonging to the unique involution of generators of R_0 determined by the pairs (Σ_0, Σ_0') and (Σ_1, Σ_1').*

The modifications of this result relevant to the cases when $\Sigma_0, ..., \Sigma_1'$ are not all distinct may be left to the reader.

The Bridging Theorem, as we have stated it, applies only when n is a multiple of 4; but we may now point out that it still holds (though less significantly) if n is of the form $4m + 2$. In fact, if $n = 4m + 2$, then q is of dimension $n - 2 = 4m$, and it cannot therefore contain any set of more than two mutually skew $[\frac{1}{2}n - 1]$'s; so that, in the second part of the theorem, the pair (Σ_1, Σ_1') must coincide with (Σ_0, Σ_0'). The proof we gave does not apply in this case because q now contains no regulus such as R_0;

but the reader should have no difficulty in supplying an alternative proof to cover this case. We may thus assert

THEOREM 3.2′. *The Bridging Theorem is valid also* (*in so far as it is applicable*) *when n is of the form* $4m + 2$.

As the final development in this section, designed to lead up to applications of the Bridging Theorem in the next section, we introduce the

DEFINITION. If n is a multiple of 4 and q is a non-singular $(n-2)$-dimensional quadric, then an irreducible system $\mathscr{Q}$ of (generally distinct) unordered pairs of generating $[\frac{1}{2}n - 1]$'s on q will be called a *regular system* if it is such that two independent generic pairs of $\mathscr{Q}$ constitute four *distinct* spaces of a regulus. (This implies that $\mathscr{Q}$ must be of positive dimension.) Further, if $\mathscr{Q}$ has the additional property that it is not contained in any more ample regular system on q, then it will be called a *maximal regular system* of pairs of generators of q.

In terms of the above definition the role of Theorems 3.1 and 3.2 in relation to the general problem of constructing linear systems of Clifford parallels can be described as follows.

Let R be a Clifford regulus in S_{2n-1}, where n is a multiple of 4, and let q be a quadric generator of the kernel K_R of R. Further let $\mathscr{Q}$ be any $(r-2)$-dimensional regular system of pairs of generating $[\frac{1}{2}n - 1]$'s of q. Then $\mathscr{Q}$ gives rise by the bridging construction to an r-dimensional irreducible system $\mathscr{F}$ of Clifford parallel $[n-1]$'s of the space S_{2n-1} containing R, of which ∞^2 arise from each pair of $\mathscr{Q}$. The system $\mathscr{F}$ so arising will not always be linear; but we now assert that if $\mathscr{Q}$ is a *maximal* regular system, then $\mathscr{F}$ is linear and maximal. For if not, then $\mathscr{F}$ would be contained (by the Embedding Theorem, Ch. 4, Th. 2.1) in a strictly larger maximal linear system $\overline{\mathscr{F}}$; and (by Th. 3.1) the spaces of $\overline{\mathscr{F}}$ would bridge the pairs of a regular system $\overline{\mathscr{Q}}$ which strictly contains $\mathscr{Q}$ – a contradiction of the assumed maximality of $\mathscr{Q}$. Moreover, if we grant the validity of the necessity clause inserted (in parentheses) in Th. 3.1 (ii), then we can add that the derived system $\mathscr{F}$, in addition to being linear and maximal, is necessarily of weight $k > 2$.

Conversely, if $\mathscr{F}$ is a maximal linear system, of weight $k > 2$, which contains R, then by Th. 3.1 the spaces of $\mathscr{F}$ bridge the pairs of a regular system $\mathscr{Q}$ on q. We assert then that this system $\mathscr{Q}$ is maximal. For if there existed a more ample regular system $\bar{\mathscr{Q}}$ containing $\mathscr{Q}$, then the spaces bridging pairs of $\bar{\mathscr{Q}}$ would form an irreducible system $\bar{\mathscr{F}}$ of Clifford parallels which (i) strictly contains $\mathscr{F}$ and (ii) is contained (by the Embedding Theorem) in a linear system $\bar{\mathscr{F}}'$; and this system $\bar{\mathscr{F}}'$ would then be a linear system strictly containing $\mathscr{F}$, in contradiction to the maximality of $\mathscr{F}$.

Recapitulating then we have

THEOREM 3.3. *When n is a multiple of 4, the existence of maximal $(r-2)$-dimensional regular systems of pairs of $[\frac{1}{2}n-1]$'s on an $(n-2)$-dimensional quadric q is equivalent to the existence of r-dimensional maximal linear systems of Clifford parallel $[n-1]$'s of S_{2n-1} of weight $k > 2$.*

For some remarks on the bridging properties of linear systems of weight 2 we refer the reader to Exs. 6 and 7 below. Systems of weight 1 (by Th. 3.1(i)) have no bridging properties.

Except for the case $n = 8$, to be discussed in the next section, we shall not pursue the study of the bridging construction any further in this book. An obvious outstanding problem, however, is that of singling out some property of a maximal regular system which will ensure that the derived linear system of Clifford parallels is non-singular.

Ex. 1. Complete the proof of the last part of Th. 3.1.

[Hint: It was supposed that

$$\{\Lambda I + S(\lambda)\}x^{(1)} = 0, \quad \{\Lambda I - S(\lambda)\}x^{(2)} = 0 \tag{1}$$

and
$$\{\Lambda' I + S(\lambda')\}(x^{(1)} + \alpha x^{(2)}) = 0. \tag{2}$$

Premultiply (2) by $S(\lambda)$ and use (1) and the extended Hurwitz–Radon matrix equations to prove that

$$\{\Lambda' I - S(\lambda')\}(x^{(1)} - \alpha x^{(2)}) = -2\phi\Lambda^{-1}(x^{(1)} + \alpha x^{(2)}). \tag{3}$$

Now solve (2) and (3) to obtain $S(\lambda')x^{(1)}$ and $S(\lambda')x^{(2)}$ and hence prove that
$$\{\Lambda' I - S(\lambda')\}(x^{(1)} + \beta x^{(2)}) = 0,$$
where $\beta = \alpha(\phi - \Lambda\Lambda')/(\phi + \Lambda\Lambda')$.]

Ex. 2. Verify that the spaces given by equations (3.7), (3.9), (3.10) and (3.11) do in fact belong to a regulus. [Hint: If A_0 and A_1 denote the respective matrices

$$\begin{pmatrix} \alpha_0 & \beta_0 \\ \gamma_0 & \delta_0 \end{pmatrix} \quad \text{and} \quad \begin{pmatrix} \alpha_1 & \beta_1 \\ \gamma_1 & \delta_1 \end{pmatrix},$$

then the matrices appearing in these equations may be written in the forms

$$A_0 \times S, \quad A_0^T \times S, \quad A_1 \times S, \quad A_1^T \times S,$$

where $\times$ denotes Kronecker product of matrices. The required result now follows from Ch. 2, § 4, Ex. 1, using the basic multiplication rule $(A \times B)(C \times D) = AC \times BD$.]

Ex. 3. State a property of a regular system $\mathscr{D}$ which will ensure that the derived bridging system of Clifford parallels is linear (cf. Cor. 3.2.3).

Ex. 4. If $\Sigma_0, \Sigma_0', \Sigma_1, \Sigma_1'$ are four distinct generators of a regulus R_0 on q, show that the *smallest* linear system of Clifford parallels which contains all spaces bridging (Σ_0, Σ_0') and all spaces bridging (Σ_1, Σ_1') must also contain all spaces bridging (Σ_2, Σ_2'), where (Σ_2, Σ_2') is any pair of the unique involution τ of generators of R_0 determined by the pairs (Σ_0, Σ_0') and (Σ_1, Σ_1'). [Hint: use Cor. 3.2.3.]

It is in fact the case (as will emerge in Ch. 6, § 2, Ex. 1) that the ∞^3-system of spaces which bridge some pair of τ is linear and non-singular – and so of weight $k = 3$. Assuming this, prove the assertion as to necessity in Th. 3.1 (ii).

Ex. 5. *Bridging construction for $n = 4$.* When $n = 4$, the bridging construction for a 4-dimensional linear system of Clifford parallel solids in S_7 must be based on a 2-dimensional regular system of pairs of generating lines of a quadric surface q. In fact the system of *all* pairs of generators of one system on q is such a regular system; and the derived construction, quite different from that projectively given in § 1, may be stated as follows:

If q is any quadric-generator of the kernel of a Clifford regulus in S_7, then all the solids of S_7 that bridge pairs of generating lines of one system on q form a (maximal) linear ∞^4-system of Clifford parallels in S_7.

Ex. 6. When n is of the form $4m + 2$, a quadric q_{n-2} cannot contain more than two mutually skew generators and so, in particular, cannot carry any regular system of pairs of generators. Also, by Th. 3.1, there can exist no linear systems of Clifford parallels of weight $k > 2$ in this case (a result in conformity with Th. A 1 of the Appendix). However, bearing in mind that we did not assume, in the Bridging Theorem, that Σ_0, Σ_0', Σ_1, Σ_1' were all distinct, it is still possible to obtain certain linear systems of weight two by a bridging construction. Namely, we consider the system of ∞^2 spaces which bridge a *fixed* pair of skew generators of q and this, as the reader may show, is linear and non-singular (cf. also Ch. 6, § 2, Ex. 1). When $n = 2$, this construction reduces to the standard one for this case given in Ch. 3, § 2, Ex. 2.

Ex. 7. An example of a linear system of weight two (when n is a multiple of 4), necessarily *not* obtainable by bridging a regular system, is the *half-fixed system* defined as follows:

Consider a system of pairs (Σ_0, Σ') of generators of q_{n-2}, in which Σ_0 is a fixed generator and Σ' varies arbitrarily in the complete system of generators of q to which Σ_0 belongs (a system of freedom $\frac{1}{8}n(n-2)$). Although these pairs do not form a regular system, the Bridging Theorem shows that the spaces bridging all such pairs form an irreducible system of Clifford parallels of freedom $\frac{1}{8}n(n-2) + 2$. The system so obtained is in fact linear of weight two. (The weight cannot be greater than two by Th. 3.3, nor less than two by the proof of Th. 3.1.) The system is, indeed, projectively equivalent to that discussed in Ch. 4, § 4, Ex. 5.

4. Clifford parallelism of [7]'s in S_{15}

By Th. 3.3 of the preceding section with $n = 8$, we know that the problem of constructing maximal linear systems, of weight $k > 2$, of Clifford parallel [7]'s in S_{15} is equivalent to that of finding maximal regular systems of pairs of solids on a 6-dimensional (non-singular) quadric q_6. We now propose to identify all possible systems of this latter kind.

The first of two preliminary results that we shall need is

PROPOSITION† 4.1. *On the Grassmannian $G(3, 7)$ of solids of S_7, the generating solids of either system on a non-singular quadric q_6 are represented by the points of a variety V_6^{128} which is the quadratic Veronesean of another non-singular quadric q_6^*.*

Proof. We choose a coordinate system in which q_6 has equation

$$x^T y \equiv x_0 y_0 + x_1 y_1 + x_2 y_2 + x_3 y_3 = 0$$

and observe that the generic solid of one system on q_6 has equation

$$y = Sx, \qquad (4.1)$$

where $S = (s_{ij})$ is a skew-symmetric matrix of order 4 whose elements above the main diagonal are independent indeterminates $s_{12}, \ldots, s_{34}$. It is easy then to check that all the Grassmann coordinates of the solid (4.1), i.e. all determinants of order 4 extracted from the 4×8 matrix $(S, -I)$, can be written as quadratic forms in $z_0, \ldots, z_7$, where

$$z_0 = 1, \quad z_1 = s_{12}, \quad z_2 = s_{13}, \quad z_3 = s_{14},$$

$$z_4 = s_{23}, \quad z_5 = s_{24}, \quad z_6 = s_{34},$$

$$z_7 = s_{12} s_{34} - s_{13} s_{24} + s_{14} s_{23};$$

and, conversely, that every quadratic form in $z_0, \ldots, z_7$ is expressible as a linear combination of the Grassmann coordinates of (4.1). This shows that the Grassmannian of the system of solids with (4.1) as generic member is the quadratic Veronesean of the quadric q_6^* which is given, in the space of the coordinates $z_0, \ldots, z_7$, by the equation

$$z_0 z_7 = z_1 z_6 - z_2 z_5 + z_3 z_4;$$

and this proves the Proposition.

The second result that we need is

PROPOSITION 4.2. *On the Grassmannian $G(3, 7)$ any regulus of solids of S_7 is represented by a rational normal quartic curve.*

The proof of this is elementary and may be left to the reader. (For a generalization, see Ch. 7, Prop. 1.2.)

---

† This proposition forms the basis of the so-called Study triality, of which we shall have more to say in Ch. 7.

Consider now the mapping of the solids of one system on q_6 – say the α-solids – by the points of q_6^*. A regulus of these solids, being represented on $G(3, 7)$ by a curve $^0C^4$ (Prop. 4.2), will be represented on q_6^* by a curve whose quadratic Veronesean is a $^0C^4$, i.e. by a conic. Hence:

Four α-solids of q_6 belong to a regulus if and only if their image points on q_6^* lie on a (proper) conic, section of q_6^* by a plane.

From this we deduce that a regular system of pairs of α-solids on q_6 must be represented on q_6^* by a system of point-pairs such that two independent generic pairs of the system are four distinct points of a conic. This implies, in particular, that the chords joining pairs of points of the system on q_{6j}^* must all intersect one another. Thus the problem of finding regular systems of pairs of α-solids on q_6 is reduced essentially to that of finding systems of lines (in the ambient [7] of q_6^*) of which any two intersect. Such a system of lines, however, must be either a star of lines or a ruled plane (or a sub-system of one or other of these). In the former case the vertex of the star must not lie on q_6^*; and in the latter case the plane must cut q_6^* in a proper conic.† We have now proved

PROPOSITION 4.3. *Every maximal regular system of pairs of α-solids on q_6 belongs to one or other of two types: (i) the type represented on q_6^* by pairs of points whose joins all pass through a fixed point O not on q_6^*, and (ii) the type represented on q_6^* by all the point-pairs of the section of q_6^* by a non-tangent plane. Systems of these two types are of dimension 6 and 2 respectively.*

A maximal regular system of the first type will be called a *Study involution* of pairs of α-generators of q_6.

From Prop. 4.3 we deduce, by applying Th. 3.3, that there exist exactly two distinct types of maximal linear system, of weight $k > 2$, of Clifford parallel [7]'s in S_{15}, namely one type of dimen-

† If, in the former case, the vertex of the star were on q_6^*, or if, in the latter case, the plane met q_6^* in a reducible conic, then the corresponding system of solids on q_6 would not be regular according to our definition. A case can be made for extending the definition of a regular system to include such specialized systems, but we shall not pursue this idea here.

sion 8 and one of dimension 4. As regards the first type – the most important since it is space-filling and non-singular (cf. Ex. 1 below) – we have

THEOREM 4.4. *If* τ *is a Study involution of pairs of generating solids of one system on any quadric-generator of the kernel of a Clifford regulus in* S_{15}, *then the* [7]*'s which bridge pairs of* τ *form a maximal linear (space-filling) system of Clifford parallels in* S_{15}.

This system is, in fact, the complex analogue of the well-known system, in real elliptic 15-dimensional space, which gives rise to the fibering of the 15-sphere by 7-spheres (cf. Ch. 1, § 5).

In the examples below we give algebraic details of each of the two types of maximal system that we have described above.

Ex. 1. *Systems of the first type.* If, in the ambient space of q_6^* (in which we took current coordinates $z_0, \ldots, z_7$) we take O to be the point $(1, 0, 0, 0, 0, 0, 0, -1)$ – not on q_6^* – we find that the associated maximal regular system of pairs of solids of q_6 has generic pair
$$x' = Sx, \quad x' = -(S^D)^{-1}x,$$

where S is a generic skew-symmetric matrix of order 4 (and S^D denotes the dual matrix, introduced in § 2). Then, using the Bridging Theorem, we may derive the equations of the associated linear ∞^8-system of Clifford parallels (after a minor notational change) in the form
$$\begin{aligned} y &= Ax' + a'x, \\ y' &= A^Dx + ax', \end{aligned}$$

where A is again skew-symmetric and a, a' are scalars (so that there are 8 parameters in all, as expected).

This, of course, is the form of the equations relative to Ω in the form $x^Tx' + y^Ty' = 0$. It is a straightforward matter to convert these equations to the canonical form employed in Ch. 4 (with Ω as $x^Ty = 0$) and we find then that the system becomes
$$y = (\alpha I + \lambda_1 S_1'' + \ldots + \lambda_7 S_7'')x,$$

where $(S_1'', \ldots, S_7'')$ is the solution of the Hurwitz–Radon matrix equations given in the Appendix, § 3 (iii). The system is consequently non-singular and (by Ch. 4, Cor. 4.2.1) space-filling of index 1.

Ex. 2. If, in the ambient space of q_6^*, we draw rays through a fixed point O lying *on* q_6^*, we thereby generate the system of pairs (Σ_0, Σ') of generators of q_6 in which Σ_0 (the generator represented by O) is fixed, and Σ' varies in the complete system of generators of q_6 to which Σ_0 belongs. By bridging this system of pairs, we obtain the half-fixed system (for $n = 8$) described in § 3, Ex. 7 and, in this sense at least, the half-fixed system may be regarded as a limiting form of the non-singular system described in Ex. 1.

Ex. 3. *Systems of the second type.* The section of q_6^* by a (non-tangent) plane represents, as we saw earlier, a regulus of generators on q_6, so that the maximal regular systems of the second type are constituted by all the (unordered) pairs of generators of a fixed regulus on q_6. Thus we may take a generic pair of the system to be given by

$$x' = \psi_0 Sx, \quad x' = \psi_0' Sx,$$

where S is now a fixed skew-symmetric matrix of order 4 and ψ_0, ψ_0' are independent indeterminates. The equation of the associated bridging system is then precisely equation (3.7) of § 3 – with α_0, β_0, γ_0, δ_0 regarded as the variable parameters.

If the system is transformed to the canonical form of Ch. 4 (with Ω as $x^T y = 0$) we find that it reduces to

$$y = (\alpha I + \lambda_1 S_1 + \lambda_2 S_2 + \lambda_3 S_3)\, x,$$

where (S_1, S_2, S_3) is an 'extra' solution of the Hurwitz–Radon equations for $n = 8$, $r = 4$, as given in the Appendix, § 3 (iii). The system is accordingly non-singular (and maximal) but not, of course, space-filling. (Cf. also Ch. 6, § 2, Ex. 1.)

Ex. 4. Any two maximal systems of the same type are projectively equivalent (by an Ω-preserving collineation in S_{15}); the two types, in fact, belong to aggregates of respective total freedoms 84 and 100.

CHAPTER 6

THE T-REPRESENTATION

In the theory of Clifford parallel spaces, as so far developed, the reader will have noted that the essential unit is not the individual $(n-1)$-dimensional space Π of S_{2n-1} but the pair of polar spaces (Π, Π') with respect to the absolute quadric Ω. Parallelism, in effect, is a relation between two such polar pairs; and Clifford reguli, as well as linear systems of Clifford parallels in general, are all systems of polar pairs. What we now propose to do in the present Chapter is to set up a birational mapping of the polar pairs (Π, Π') of S_{2n-1} on the points of a certain algebraic variety T, of dimension n^2; and this mapping – the T-representation – will be such that the polar pairs belonging to any Clifford regulus in S_{2n-1} are mapped on T by the points of a line. It will then follow that the polar pairs of any linear system of Clifford parallels in S_{2n-1} are mapped on T by the points of a flat space. The T-representation which we define in this way is different from other representations of the polar pairs (Π, Π') of S_{2n-1} – to be discussed in Ch. 7 – which arise from $(2, 1)$-projections of the Grassmannian $G(n-1, 2n-1)$.

1. Basis of the representation

The idea of the T-representation is to associate with any general pair (Π, Π') of polar $[n-1]$'s with respect to Ω a unique quadric primal $\mathscr{Q}(\Pi, \Pi')$, and then to take the coefficients in the equation of $\mathscr{Q}(\Pi, \Pi')$ as the coordinates of the pair.

More precisely, supposing that Π and Π' are chordal, we denote by q, q' the $(n-2)$-dimensional quadrics in which Π and Π' meet Ω, and we consider the quadric $[n-1]$-cones $\Pi(q')$ and $\Pi'(q)$ that join Π to q' and Π' to q. Each of these cones meets Ω in the quartic $(2n-3)$-dimensional variety generated by all the lines joining points of q to points of q', so that the pencil of quadrics defined by $\Pi(q')$ and $\Pi'(q)$ contains Ω. Within this pencil, then, there is a unique quadric $\mathcal{Q}(\Pi, \Pi')$ which is the harmonic conjugate of Ω with respect to $\Pi(q')$ and $\Pi'(q)$, and this is the quadric that we propose to associate with the pair (Π, Π'). The birational character of the correspondence between polar pairs (Π, Π') and quadrics $\mathcal{Q}(\Pi, \Pi')$ follows at once from the observation that the pencil defined by $\Pi(q')$ and $\Pi'(q)$ contains no other cones.

To identify $\mathcal{Q}(\Pi, \Pi')$ algebraically, we choose the coordinate system so that Ω has equation $x^T y = 0$; and we suppose then, as we may, that Π and Π' have equations $y = Ax$ and $y = -A^T x$, where A is a matrix of order n. A straightforward calculation then gives the equations of the cones $\Pi(q')$ and $\Pi'(q)$ in the respective forms

$$K_1 \equiv (y - Ax)^T (A + A^T)^{-1} (y - Ax) = 0,$$

$$K_2 \equiv (y + A^T x)^T (A + A^T)^{-1} (y + A^T x) = 0,$$

provided that $A + A^T$ is non-singular, i.e. that Π is skew to Π'. We now write $(A + A^T)^{-1} = B = B^T$, so that

$$K_1 \equiv y^T By - x^T A^T By - y^T BAx + x^T A^T BAx,$$

$$K_2 \equiv y^T By + x^T ABy + y^T BA^T x + x^T ABA^T x.$$

By the definition of B

$$BA + BA^T = AB + A^T B = I,$$

i.e. $$BA^T = I - BA \quad \text{and} \quad A^T B = I - AB$$

so that $$ABA^T = A^T BA \quad (= A - ABA).$$
Hence

$$K_2 - K_1 \equiv x^T (A + A^T) By + y^T B(A + A^T) x \equiv 2x^T y,$$

and this exhibits Ω as a member of the pencil given by

$$K_2 + \lambda K_1 = 0.$$

Now $\mathcal{Q}(\Pi, \Pi')$ was defined to be the harmonic conjugate of Ω with respect to the two cones $K_1 = 0$, $K_2 = 0$ of the pencil, so that its equation is $K_2 + K_1 = 0$; and this, after some reduction, gives the equation of $\mathcal{Q}(\Pi, \Pi')$ in the form

$$x^T A (A + A^T)^{-1} A^T x + x^T (A - A^T)(A + A^T)^{-1} y$$
$$+ y^T (A + A^T)^{-1} y = 0. \quad (1.1)$$

We take the coefficients in this equation, i.e. effectively the elements of the matrix triad

$$[A(A + A^T)^{-1} A^T, (A - A^T)(A + A^T)^{-1}, (A + A^T)^{-1}], \quad (1.2)$$

to be the *coordinates of the polar pair* (Π, Π')—the coordinates, in other words, of the image point of $\mathcal{Q}(\Pi, \Pi')$ in the space S_N ($N = \frac{1}{2}(2n-1)(2n+2)$) of quadrics of S_{2n-1}. If the elements a_{ij} of A are taken to be independent indeterminates, then the elements of (1.2) are the coordinates of the generic point of an *image variety* T of the polar pairs (Π, Π'). It should be noted, however, that although (1.2) exhibits T as a rational transform of the (affine) space S_{n^2} of the a_{ij}, the associated correspondence between T and S_{n^2} is one–two, since the same point of T arises equally from A and $-A^T$. We shall also have to consider the fundamental elements in the mapping of the pairs (Π, Π') by points of T, such as arise when Π and Π' are not skew to one another, so that $A + A^T$ is singular.

The variety T, as above defined, lies in a prime S_{N-1} of S_N. For the pencil of quadrics $K_2 + \lambda K_1 = 0$ has the property that any two of its members which separate the base cones $K_1 = 0$ and $K_2 = 0$ harmonically are mutually apolar; and indeed each of them is self-reciprocal with respect to the other.† Thus $\mathcal{Q}(\Pi, \Pi')$ is always apolar to Ω, so that its image point in S_N lies in a fixed prime S_{N-1} (cf. Ex. 1 below).

Note. There are other ways of constructing the variety T. Thus, for example, if $\mathcal{T}$ is the image variety of all the quadric cones $\Pi(q')$, i.e. all the tangent cones to Ω from spaces Π, and if O is the image point of Ω in S_N, then T is the (2,1) projection of $\mathcal{T}$ from O onto a prime S_{N-1}. Alternatively we may define T

---

† This follows from the observation that a coordinate system can be chosen in S_{2n-1}, with Π, Π' as reference spaces X, Y, such that the cones and any pair of quadrics separating them harmonically have equations $x^T x = 0$, $y^T y = 0$ and $x^T x \pm y^T y = 0$.

as the model of all the *pencils* of quadrics such as $K_2 + \lambda K_1 = 0$. See also Ex. 3 below.

We now prove the following basic property of the T-representation.

THEOREM 1.1. *On the T-model as above defined, every Clifford regulus of S_{2n-1} is represented on T by a line;[†] and every linear system of Clifford parallels is represented on T by a linear space.*

Proof. Let R_0 be any Clifford regulus of S_{2n-1} and let the coordinate system be chosen so that Ω has equation $x^T y = 0$ and R_0 is given by $y = \lambda x$. Then the equations of any pair (Π, Π') of polar spaces of R_0 are of the form $y = \pm \lambda x (\lambda \neq 0)$. On making the substitution $A = \lambda I$ in (1.1) we obtain the equation of $\mathcal{Q}(\Pi, \Pi')$ in the form

$$\lambda^2 x^T x + y^T y = 0;$$

and, since the coefficients in this equation are linear in the single parameter λ^2, it follows that R_0 is represented on T by a line. Further, since our definition of the T-model was independent of any choice of the coordinate system in S_{2n-1}, it follows that *every* Clifford regulus is represented on T by a line.

To see now, further, that every linear system $\mathcal{F}$ of Clifford parallels in S_{2n-1} is represented on T by a linear space, it is only necessary to point out that $\mathcal{F}$ must be represented on T by an irreducible variety $\mathcal{V}$ with the property that $\mathcal{V}$ contains the line joining a generic pair of points of itself, i.e. $\mathcal{V}$ is a flat space. This completes the proof.

Ex. 1. Show that the linear relation (apolarity to Ω) satisfied by all the quadrics $\mathcal{Q}(\Pi, \Pi')$ is reflected in the fact that the matrix $(A - A^T)(A + A^T)^{-1}$ has zero trace.

Ex. 2. Show that the 'dual' T-model, obtained by assigning to (Π, Π') the coefficients in the tangential equation of $\mathcal{Q}(\Pi, \Pi')$, is projectively equivalent to the T-model. Deduce that the quadrics associated with the polar-pairs (Π, Π') of a linear system of Clifford parallels form a linear system of quadrics which is also a linear system of quadric envelopes.

---

† This result explains why it is natural to say that a Clifford regulus is the 'join' of any two of its generators.

Ex. 3. If Π and Π' (as polar $[n-1]$'s with respect to Ω) have equations $y = Ax$ and $y = -A^Tx$, show that the harmonic biaxial involution with Π and Π' as axes is given by

$$\begin{pmatrix} x \\ y \end{pmatrix} \rightarrow \begin{pmatrix} (A+A^T)^{-1}(A^T-A) & 2(A+A^T)^{-1} \\ 2A(A+A^T)^{-1}A^T & (A-A^T)(A+A^T)^{-1} \end{pmatrix} \begin{pmatrix} x \\ y \end{pmatrix}$$

and deduce from this an alternative method of defining the T-model.

Ex. 4. *The T-models for $n = 2$ and 3.* We give here some preliminary details about these two models in advance of their further consideration in § 3.

(i) For $n = 2$, we find easily, from the parametric equations (1.2), that the T-model is a $V_4^6[8]$, Segre product of two planes, lying in a prime of the image space S_9 of quadrics of S_3. The two kinds of linear ∞^2-systems of Clifford parallel lines of S_3 are represented on T by the planes of its two generating systems.

(ii) For $n = 3$, we may identify the T-model as follows. If Ω in S_5 has equation $x_0y_0 + x_1y_1 + x_2y_2 = 0$, then the coordinates $(x_0, x_1, x_2, y_0, y_1, y_2)$ of a point of Ω can be regarded as those of a line of S_3; and a pair of polar planes (Π, Π') with respect to Ω meet Ω in conics q, q' which represent complementary reguli on a quadric ψ of S_3. Further, the complex of tangent lines to ψ (the quadric complex determined by ψ) corresponds to the quartic threefold on Ω generated by all the lines meeting q and q'; and this threefold is the base of a pencil of quadrics which contains Ω. Then $\mathcal{Q}(\Pi, \Pi')$, being the unique quadric of this pencil which is apolar to Ω, is seen to be that given (essentially) by the *standard* equation of the quadric complex determined by ψ, i.e. that in which the coefficients are all the minors of order two from the matrix of ψ. The generic point of the resulting T-model is accordingly that whose coordinates in S_{19} are all the minors of order two extracted from a symmetric matrix of order four with independent indeterminates as elements; and, from this form of its generic point, the T-model is immediately identified as the known† variety $V_9^{92}[19]$.

† This variety has been discussed in detail by Todd[26] in connection with his work on the Cayley model of conics in S_3, and further properties of it were given by Semple[21] in his investigations of complete quadrics.

2. The T-representation of augmented generators

As we have already remarked, the formulae (1.2) which give the coordinates of the image point of a polar pair (Π, Π') no longer apply when Π meets Π', the matrix $A + A^T$ being then singular. We shall return later to the general problems arising from this; but for the moment we are concerned with the extreme case in which Π' coincides with Π, i.e. with a generator α or β of Ω which counts as a coincident polar pair (α, α) or (β, β). What we now wish to show is that whereas such a coincident pair, say (α, α), is fundamental for the transformation, being such that its total image in T is a linear space A of dimension $\frac{1}{2}(n-1)(n+2)$, each *augmented generator* $\alpha(q)$ based on α (cf. Ch. 3, §5) corresponds in a well-defined manner to its own particular point of A. Each $\alpha(q)$, in fact, where q is any nonsingular quadric in α, functions as a specialized limit of a polar pair (Π, Π') as Π and Π' both approach α within a Clifford regulus which has α as one of its terminal generators and q as the corresponding terminal quadric; and to this specialized limit there corresponds a definite point of A. More precisely we prove

PROPOSITION 2.1. *In the birational mapping of the polar pairs* (Π, Π') *of* S_{2n-1} *on the points of* T, *each generator* α *or* β *of* Ω *(regarded as a coincident polar pair) is fundamental, having a space* A *or* B, *of dimension* $\frac{1}{2}(n-1)(n+2)$, *as its total image on* T. *Any general point of* A *corresponds in a well-defined way to a unique augmented generator* $\alpha(q)$; *and similarly for* B. *As* α *varies, the image points of all the augmented generators of the type* $\alpha(q)$ *lie on a primal* $\mathscr{A}$ *of* T, *the* $(n^2 - 1)$-*dimensional subvariety of* T *generated by the* $\infty^{\frac{1}{2}(n^2-n)}$ *spaces* A; *and similarly the image points of all the* $\beta(q)$ *lie on a primal* $\mathscr{B}$ *of* T, *generated by the spaces* B. *We call* $\mathscr{A}$ *and* $\mathscr{B}$ *the* nuclear primals *of* T.

Proof. We first make two observations:

(i) In Ch. 3, §5 we said that two augmented generators $\Pi_1(q_1)$ and $\Pi_2(q_2)$ were *compatible* if they belonged to the same Clifford regulus R, i.e. if Π_1 and Π_2 were the terminal generators and q_1, q_2 the terminal quadrics of R. Also we noted (Ch. 3, Prop. 5.3) that if γ_1, γ_2 are any two skew generators of Ω, and if $\gamma_1(q_1)$ is an augmentation of γ_1, then there exists a unique augmentation $\gamma_2(q_2)$ of γ_2 that is compatible with $\gamma_1(q_1)$.

(ii) As regards the parametrization of T by elements of the three matrices (1.2), we observe that this can be simplified by writing $A = C + S$, where C is symmetric and S is skew-symmetric. We then have $A^T = C - S$; whence, after removal of a factor of proportionality, the triad of matrices takes the form

$$[C - SC^{-1}S, \ 2SC^{-1}, \ C^{-1}]. \tag{2.1}$$

The original n^2 indeterminates in A are now replaced by $\frac{1}{2}(n^2 + n)$ new indeterminates in C and $\frac{1}{2}(n^2 - n)$ in S. We note that (C, S) and $(-C, S)$ give the same point of T.

For the proof of Prop. 2.1 we may obviously confine our attention to one of the two types of augmented generator, say $\alpha(q)$.

Suppose then that Ω has equation $x^T y = 0$ and that α is the generator of Ω given by $y = Sx$, where S is skew-symmetric. This implies, incidentally, that α is skew to the reference space Y.

If D is any non-singular symmetric matrix, then the cone $y^T D y = 0$ meets Y in a non-singular quadric q^* and thereby defines an augmented generator $Y(q^*)$ attached to Y. By (i) above, therefore, there exists a unique compatible augmented generator $\alpha(q)$ attached to α, and joined therefore to $Y(q^*)$ by a Clifford regulus R. We assert then that the equation of R is

$$y = (S + tD^{-1})x, \tag{2.2}$$

where t is the variable parameter. For plainly the regulus given by (2.2) contains Y and α; it is autopolar (and therefore Clifford), polar pairs in R being given by equal and opposite values of the parameter t; and it is easy to verify that its terminal quadric in Y is q^*. Moreover, its terminal quadric q in α is given by $x^T D^{-1} x = 0$. We note also, conversely, that every augmented generator attached to α is terminal in a unique Clifford regulus of the form (2.2).

To find the image point of $\alpha(q)$ on T, we replace C in (2.1) by tD^{-1} and then make t tend to zero; and, after removal of a factor of proportionality, we find that *the required image point (of the augmented generator $\alpha(q)$ given by $y = Sx$ with augmentation*

$x^T D^{-1} x = 0$) *is that whose coordinates are the elements of the matrices of the triad* $[-SDS, 2SD, D]$. (2.3)

This image point depends therefore, in all, on the values of $n^2 - 1$ parameters, namely on the $\frac{1}{2}(n^2 - n)$ elements s_{ij} $(i < j)$ in S and on the $\frac{1}{2}(n^2 + n) - 1$ *ratios* of the elements $d_{ij} = d_{ji}$ of D.

If we keep the s_{ij} fixed, then the elements of the first and second of the three matrices (2.3) are only linear combinations of those of the third; and this means that all the augmented generators $\alpha(q)$ attached to a fixed α have image points which lie in a space A of dimension $\frac{1}{2}(n-1)(n+2)$ on T. If the s_{ij} are now also allowed to vary independently, then we see that the image points of all the $\alpha(q)$, for variable α, lie on a primal $\mathscr{A}$ of T, locus of the $\infty^{\frac{1}{2}(n^2 - n)}$ spaces A, and similarly the image points of all the $\beta(q)$, for variable β, lie on another primal $\mathscr{B}$ of T. This completes the proof.

If $L(\mathscr{F})$ is the space on T which represents a linear system $\mathscr{F}$, then the augmented generators of $\mathscr{F}$ will be represented on $L(\mathscr{F})$ by points of its intersection with $\mathscr{A}$ or $\mathscr{B}$. The following Proposition describes the nature of this intersection.

PROPOSITION 2.2. *If* $L(\mathscr{F})$ *is the* r-*dimensional space on* T *which maps the polar pairs of an* r-*dimensional linear system* $\mathscr{F}$ *of Clifford parallels in* S_{2n-1}, *then the points of* $L(\mathscr{F})$ *which map the augmented generators belonging to* $\mathscr{F}$ *lie on a quadric primal* $\mathscr{N}(\mathscr{F})$ *of* $L(\mathscr{F})$ – *the* nuclear quadric *of the mapping of* $\mathscr{F}$ *on* $L(\mathscr{F})$.

Proof. Let $\mathscr{F}$ be given by the canonical equation

$$y = (\alpha I + \lambda_1 S_1 + \ldots + \lambda_{r-1} S_{r-1}) x,$$

where the S_i satisfy the equations (3.1) of Ch. 4 for some value of k $(1 \leqslant k \leqslant r)$. For the relevant properties of such canonical equations we refer the reader to Ch. 4, § 4, and we add here the following remark:

(A) If $\lambda_1, \ldots, \lambda_{r-1}$ are regarded as independent indeterminates, then each of them can be expressed (though not necessarily uniquely) as a linear combination with constant coefficients of the elements of the matrix $\lambda_1 S_1 + \ldots + \lambda_{r-1} S_{r-1}$. (For otherwise the linear independence of $S_1, \ldots, S_{r-1}$ would be contradicted.)

To find the image point on T of the general polar pair (Π, Π') of $\mathscr{F}$ we substitute $A = \pm \alpha I + \lambda_1 S_1 + \ldots + \lambda_{r-1} S_{r-1}$ in (1.2) and make use of the relations (3.1) of Ch. 4 to simplify the result. We find then that the coordinates of the required image point are the elements of the matrices of the triad

$$[(\alpha^2 + \lambda_1^2 + \ldots + \lambda_{k-1}^2) I, \; 2\lambda_1 S_1 + \ldots + 2\lambda_{r-1} S_{r-1}, \; I].$$

By (A) above we can then make an allowable transformation to new coordinates $\xi_0, \ldots, \xi_N$ in the space of T so that the image point of (Π, Π') is given by

$$\left. \begin{array}{l} \xi_0 = 1, \quad \xi_1 = \lambda_1, \ldots, \quad \xi_{r-1} = \lambda_{r-1}, \\ \xi_r = \alpha^2 + \lambda_1^2 + \ldots + \lambda_{k-1}^2, \quad \xi_{r+1} = \ldots = \xi_N = 0. \end{array} \right\} \quad (2.4)$$

For variable $\alpha, \lambda_1, \ldots, \lambda_{r-1}$, these are, as expected, the parametric equations of a linear space $L(\mathscr{F})$ on T. But now, putting $\alpha = 0$, we see that the subvariety in $L(\mathscr{F})$ which maps the augmented generators of $\mathscr{F}$ is the quadric $\mathscr{N}(\mathscr{F})$ given by

$$\xi_0 \xi_r = \xi_1^2 + \ldots + \xi_{k-1}^2, \quad \xi_{r+1} = \ldots = \xi_N = 0.$$

This completes the proof; but it should be noted that our analysis does not deal explicitly with those spaces of $\mathscr{F}$ which meet the reference space Y and which can only be represented in $L(\mathscr{F})$ by points which lie in the 'prime at infinity' of this space, given by $\xi_0 = 0$. We shall return to this later in §5.

COROLLARY 2.2.1. *The nuclear quadric $\mathscr{N}(\mathscr{F})$ is non-singular if and only if $\mathscr{F}$ is non-singular $(k = r)$. Otherwise, if $k < r$, $\mathscr{N}(\mathscr{F})$ is a quadric cone with an $[r-k-1]$ as vertex; and in particular, if $\mathscr{F}$ is totally singular $(k = 1)$, $\mathscr{N}(\mathscr{F})$ breaks up into the prime at infinity in $L(\mathscr{F})$, given by $\xi_0 = 0$, together with another prime.*

This Corollary, it will be noted, provides a natural geometrical interpretation of the weight k of $\mathscr{F}$ and shows, in particular, that k is an invariant of $\mathscr{F}$ (cf. Ch. 4, §4).

A Clifford regulus R within $\mathscr{F}$, as already noted (Th. 1.1), is mapped on $L(\mathscr{F})$ by a line, and we see now, from Prop. 2.2, that the terminal (augmented) generators of R correspond to the two points (necessarily distinct) in which this line meets the nuclear quadric $\mathcal{N}(\mathcal{F})$; in §5 we shall see, conversely, that *every* proper chord of $\mathcal{N}(\mathcal{F})$ represents a Clifford regulus within $\mathcal{F}$. (A line which touches $\mathcal{N}(\mathcal{F})$ represents what may be called a *split* Clifford regulus, as described in Ex. 2 below.)

If $\mathcal{F}$ is singular, so that $\mathcal{N}(\mathcal{F})$ is a cone, then any space of $\mathcal{F}$ that is mapped on a point of the vertex of $\mathcal{N}(\mathcal{F})$ can clearly belong to no Clifford regulus within $\mathcal{F}$ and is therefore a singular space of $\mathcal{F}$ (cf. Ch. 4, Cor. 4.1.1).

As a final remark, we mention that there can exist what we may call *degenerate linear systems* $\mathcal{F}^*$ in S_{2n-1}, such that the spaces of a system $\mathcal{F}^*$ are mapped on T by the points of a flat space $L(\mathcal{F}^*)$ in which the nuclear quadric is a repeated prime. But since such systems $\mathcal{F}^*$ contain no Clifford reguli, they are of little geometric interest.

Ex. 1. Discuss the representation on the T-model of the system $\mathcal{G}$ of ∞^4 spaces given by equation (3.7) of Ch. 5, §3, where S is a fixed non-singular skew-symmetric matrix of order $\tfrac{1}{2}n$ and $\alpha_0, \ldots, \delta_0$ are independent parameters. (It should be remembered that in the above context the equation of Ω was $x^T x' + y^T y' = 0$.) The coordinate ratios of the image point on T of a space of $\mathcal{G}$ all turn out to be linear combinations of

$$z_0 = 1, \ z_1 = \alpha_0, \ z_2 = \delta_0, \ z_3 = \beta_0 + \gamma_0 \ \text{and} \ z_4 = \alpha_0 \delta_0 - \beta_0 \gamma_0,$$

so that $\mathcal{G}$ is a *linear* system of dimension 4. The augmented generators that belong to $\mathcal{G}$ occur when $\beta_0 = \gamma_0$, and they are therefore represented on T by points of the nuclear quadric $\mathcal{N}(\mathcal{G})$ with equation

$$z_3^2 - 4z_1 z_2 + 4z_0 z_4 = 0.$$

As this quadric is non-singular, it follows that $\mathcal{G}$ is also non-singular.

Show further that if the eigenvalues of

$$\begin{pmatrix} \alpha_0 & \beta_0 \\ \gamma_0 & \delta_0 \end{pmatrix}$$

are required to be mates in a fixed involution, then the corresponding spaces of $\mathcal{G}$ are represented on T by the points of a solid which does not touch $\mathcal{N}(\mathcal{G})$. These spaces therefore form a non-singular ∞^3 linear subsystem of $\mathscr{G}$, i.e. a system of weight 3. Interpret this result in the context of Ch. 5, § 3, with particular reference to Ex. 4 there following; and hence complete the proof of Th. 3.1 of Ch. 5.

Show finally that the spaces of $\mathscr{G}$ corresponding to matrices

$$\begin{pmatrix} \alpha_0 & \beta_0 \\ \gamma_0 & \delta_0 \end{pmatrix}$$

with two fixed unequal eigenvalues form a non-singular linear system of dimension 2.

Ex. 2. *Split Clifford reguli.* Let $\mathscr{F}$ be a linear ∞^r-system of Clifford parallels $(r > 1)$ given by the canonical equation (3.2) of Ch. 4, let $\lambda_1, \ldots, \lambda_{r-1}$ be independent indeterminates and, as usual, let $\Lambda^2 = -(\lambda_1^2 + \ldots + \lambda_{k-1}^2)$, where k is the weight of $\mathscr{F}$. Suppose also that $k > 1$, so that $\Lambda \neq 0$ and n is necessarily even. Then the spaces $\Pi(t)$, $\Pi'(t)$ of $\mathscr{F}$ given respectively by

$$y = t(\Lambda I + \lambda_1 S_1 + \ldots + \lambda_{r-1} S_{r-1}) x,$$

$$y = t(-\Lambda I + \lambda_1 S_1 + \ldots + \lambda_{r-1} S_{r-1}) x,$$

are a pair of polar spaces with respect to Ω represented on the T-model (in the notation of equation (2.4)) by the point

$$\xi_0 = 1, \; \xi_1 = t\lambda_1, \ldots, \; \xi_{r-1} = t\lambda_{r-1}, \; \xi_r = 0, \; \xi_{r+1} = \ldots = \xi_N = 0$$

and this point, as t varies, describes a generic tangent line to the nuclear quadric $\mathcal{N}(\mathscr{F})$ at the point $(1, 0, \ldots, 0)$.

The set of pairs $\Pi(t)$, $\Pi'(t)$, for given $\lambda_1, \ldots, \lambda_{r-1}$ and variable t, must therefore be regarded as a limiting kind of Clifford regulus – which we referred to above as a *split Clifford regulus* – whose terminal spaces have coincided. As an algebraic variety, such a split regulus has broken up into two halves ρ and ρ', each of dimension n and order $\frac{1}{2}n$, ρ being generated (for variable t) by $\Pi(t)$ and ρ' by $\Pi'(t)$. We shall not refer further to split reguli in this book and content ourselves with the following brief notes.

(i) The generators of ρ are the polars of those of ρ'.

(ii) The two halves have a generator of Ω as a common generator, and ρ and ρ' have no other common points. The common generator, like the terminal generators of a proper Clifford regulus, has a natural structure of an augmented generator.

(iii) Each half is a cone generated by the $[n-1]$'s joining a vertex $[\frac{1}{2}n-1]$ to an ordinary regulus of $[\frac{1}{2}n-1]$'s and lies therefore in a $[\frac{3}{2}n-1]$. The common generator mentioned in (ii) is the join of the vertices of these cones or, equally, the intersection of their ambient spaces.

For $n=2$, let g be a generator of the absolute quadric surface Ω, let A, B be two points of g and let α, β be the tangent planes to Ω at A, B respectively. Then the pencils of lines $A(\beta)$ and $B(\alpha)$ are the two halves of a split Clifford regulus in S_3, the common (augmented) generator being $g(A, B)$.

Ex. 3. Show that the Clifford reguli that are contained in a linear system $\mathscr{F}$, given by its canonical equation in the usual form, are represented in the affine space of coordinates $\alpha, \lambda_1, ..., \lambda_{r-1}$ by conics of a certain family of which a general member has equations of the form

$$\alpha^2 + \lambda_1^2 + ... + \lambda_{k-1}^2 = a_1\lambda_1 + b_1 = ... = a_{r-1}\lambda_{r-1} + b_{r-1},$$

where the a_i and b_j are arbitrary constants. Discuss the character of this family for low values of r.

3. Geometrical construction of T

Apart from its connection with the theory of Clifford parallels, the variety T is of some interest on its own account as a model of polar pairs (Π, Π') with respect to the quadric Ω. It would seem worth while, therefore, to investigate its geometrical character and construction more closely, and this and the following section are devoted to a digression on this topic. We shall show in this section that T is uniquely defined by a certain subvariety $\mathscr{V}$ of itself – its core variety – and that $\mathscr{V}$ is the quadratic Veronesean of a non-singular quadric Ω'_{2n-2}.

We begin by recalling from the previous section that if α is the generator $y = Sx$ of Ω, and if q is the quadric in which α is met by the cone $x^T D^{-1} x = 0$, where the matrix D is symmetric and non-singular, then the coordinates of the image point on T of the augmented generator $\alpha(q)$ are the elements of the matrices of the triad

$$[-SDS, \ 2SD, \ D]. \tag{3.1}$$

If the elements $s_{ij} = -s_{ji}$ of S and the elements $d_{ij} = d_{ji}$ of D are all taken to be independent indeterminates, then (3.1) can be regarded as the generic point of the nuclear primal $\mathscr{A}$ which is the image variety on T of all $\alpha(q)$; and $\mathscr{A}$ is generated by spaces A on T which each correspond to a replacement of the s_{ij} in (3.1) by a set of constants which are not all zero. Similarly, there exists on T a nuclear primal $\mathscr{B}$ which is the image variety of all $\beta(q)$; and the two primals $\mathscr{A}$ and $\mathscr{B}$ are generated respectively by the systems of spaces A and B on T which correspond to the individual generators α and β of Ω.

We now specialize D in (3.1) by taking it to be of rank 1, writing $D = \xi\xi^T$, where ξ is a column vector of indeterminates. Then (3.1) becomes

$$[-S\xi\xi^T S, \ 2S\xi\xi^T, \ \xi\xi^T]; \tag{3.2}$$

and if we write $\eta = S\xi$, this takes the form

$$[\eta\eta^T, \ 2\eta\xi^T, \ \xi\xi^T], \tag{3.3}$$

where ξ and η are subject only to the fixed relation

$$\xi^T\eta = 0. \tag{3.4}$$

If then we let Ω' be the quadric primal $x'^T y' = 0$ of a space S'_{2n-1} in which (x', y') are the current coordinates, we see that (ξ, η) is a generic point of Ω' and hence that (3.3) – subject to (3.4) – is a generic point of a subvariety $\mathscr{V}$ of T which is the quadratic Veronesean transform of Ω'. (To avoid confusion, we do not identify S'_{2n-1} and Ω' with S_{2n-1} and Ω; in fact, the matrix triad (3.3) is the coefficient image of the quadric

$$(x^T\eta + y^T\xi)(\eta^T x + \xi^T y) = 0$$

– an extreme specialization of the quadric $\mathscr{Q}(\Pi, \Pi')$ given by (1.1) – which consists of the repeated tangent prime to Ω at (ξ, η); thus the points of Ω' (or $\mathscr{V}$) are in natural correspondence with the tangent primes of Ω, so that Ω' is essentially the dual of Ω.) We shall refer to the subvariety $\mathscr{V}$ as the *core* of T, postponing till the following section its interpretation in relation to the T-representation.

The intersection of $\mathscr{V}$ with a generator A of $\mathscr{A}$, obtained by specializing S in (3.2) to a constant matrix, is a variety ϕ, of dimension $n-1$, which is the quadric Veronesean of an $[n-1]$-generator of one system of Ω' – say an α'-generator; and plainly A, being of dimension $\frac{1}{2}(n-1)(n+2)$, is the ambient space of ϕ. Thus $\mathscr{A}$ is now identified with the locus of ambient spaces A of the quadric Veroneseans ϕ on $\mathscr{V}$ which correspond to the α'-generators of Ω'. It follows now, by symmetry, that $\mathscr{B}$ is generated by the ambient spaces B of the quadric Veroneseans ψ on $\mathscr{V}$ which correspond to the β'-generators of Ω'. Hence:

(i) *The model T possesses a* core variety $\mathscr{V}$, *of dimension $2n-2$, which is the quadratic Veronesean of a non-singular quadric Ω' of S'_{2n-1}. The primals $\mathscr{A}$ and $\mathscr{B}$ of T (arising from the systems of $\alpha(q)$ and $\beta(q)$ of S_{2n-1}) are the loci of the ambient spaces A and B of the two systems of quadric Veroneseans ϕ and ψ on $\mathscr{V}$ which correspond to generators α' and β' of Ω'.*

It now remains to be shown that the core variety $\mathscr{V}$ uniquely determines not only $\mathscr{A}$ and $\mathscr{B}$ but also T itself. We propose to do this by showing that $\mathscr{V}$ of itself determines the whole family of lines of T (generating T multiply) which represent Clifford reguli of S_{2n-1}.

The argument is the same whether n is even or odd; in other words, whether the terminal generators of a Clifford regulus are of the same or of opposite systems on Ω. *We shall suppose, therefore, for convenience, that n is even.*

With this proviso, we consider a pair of skew generators α_1, α_2 of Ω and the corresponding generating spaces A_1 and A_2 of $\mathscr{A}$. A pair of points Q_1 and Q_2 of A_1 and A_2 respectively represent in general a pair of augmented generators $\alpha_1(q_1)$ and $\alpha_2(q_2)$; and the line Q_1Q_2 lies on T and represents a Clifford regulus if and only if $\alpha_1(q_1)$ and $\alpha_2(q_2)$ are compatible (cf. Ch. 3, § 5). Further, if α_1 and α_2 are fixed, the condition of compatibility implies that each of q_1, q_2 uniquely determines the other. There exists, therefore, a Cremona transformation between points Q_1, Q_2 of A_1 and A_2 respectively which represent compatible augmented generators $\alpha_1(q_1)$ and $\alpha_2(q_2)$.

To see the nature of this correspondence, we take α_1 and α_2 to be the generators X and Y of Ω. In this case, if q_2 in Y has equation $y^T D y = 0$, then its compatible q_1 in X has equation $x^T D' x = 0$

where $D' = D^{-1}$. The spaces A_1 and A_2 are those defined by the matrix triads $[0, 0, D]$ and $[D', 0, 0]$; and the Veroneseans ϕ_1 and ϕ_2 in which A_1 and A_2 meet $\mathscr{V}$ correspond to the generators $y' = 0$ and $x' = 0$ of Ω', their parametric representations being $(0, 0, \xi\xi^T)$ and $(\eta\eta^T, 0, 0)$ respectively. Then we may say:

(ii) *There exists, between points $Q_1 = (d_{ij})$ of A_1 and $Q_2 = (d'_{ij})$ of A_2, a symmetric determinantal Cremona transformation,† with matrix equation $D' = D^{-1}$, such that the lines Q_1Q_2 which join corresponding points lie on T and represent Clifford reguli of S_{2n-1}.*

Further, if $n > 2$, the key base varieties of the above transformation are the Veroneseans ϕ_1 and ϕ_2 in which A_1 and A_2 meet $\mathscr{V}$.

We shall suppose from now on that $n > 2$. The case $n = 2$, when the Cremona transformation is a plane collineation, is discussed separately below.

A Cremona transformation of the type described above, between spaces A_1 and A_2 which are each of dimension $\frac{1}{2}(n^2 + n) - 1$, will be called a $\tau(n)$; that is to say, the transformation must be such that, for suitable choice of the symmetrical arrays of coordinates d_{ij} in A_1 and d'_{ij} in A_2, it can be represented by the matrix equation $D' = D^{-1}$.

Suppose now that only the key base varieties ϕ_1 and ϕ_2 of such a $\tau(n)$ are prescribed, and that these have parametric representations, relative to fixed coordinate systems in A_1 and A_2, given by $(d_{ij}) = \xi\xi^T$ and $(d'_{ij}) = \eta\eta^T$. Then, if τ_0 is the transformation given by $D' = D^{-1}$, the equation of the general $\tau(n)$ with ϕ_1 and ϕ_2 as its key base varieties is of the form

$$MD'M^T = D^{-1}, \tag{3.5}$$

where M is any non-singular $n \times n$ matrix of constants; for this equation can be written in the form $D'' = D^{-1}$ by use of the coordinate transformation $D'' = MD'M^T$ in A_2, and it represents the product of τ_0 by a self-collineation of A_2 which leaves ϕ_2 invariant.

† This is, of course, the well-known Cremona transformation between the space of quadric loci of $[n-1]$ and the space of the corresponding quadric envelopes. The key base varieties are those representing the repeated primes of $[n-1]$ and the repeated points of $[n-1]$ respectively.

Now, for any $\tau(n)$ of the above kind, the neighbourhood of a point ξ of ϕ_1 is transformed into the ambient space of a sub-Veronesean variety ρ_2 of ϕ_2 – image of an $[n-2]$ of the $[n-1]$ of which ϕ_2 is the Veronesean; and similarly the neighbourhood of a point η of ϕ_2 is transformed into the ambient space of a sub-Veronesean ρ_1 of ϕ_1. It is easy to see then that both correspondences – from ξ to ρ_2 and from η to ρ_1 – arise from one and the same *correlation*† between ϕ_1 and ϕ_2; and that, for the $\tau(n)$ given by (3.5), the equation of this correlation is

$$\xi^T M \eta = 0.$$

Thus when ϕ_1 and ϕ_2 are assigned, a transformation $\tau(n)$ between A_1 and A_2, with ϕ_1 and ϕ_2 as its key base varieties, is uniquely determined by a non-singular correlation between ϕ_1 and ϕ_2 – that which determines the fundamental spaces in A_1 and A_2 which correspond to the neighbourhoods of points of ϕ_2 and ϕ_1. In particular, the correlation which determines τ_0 in this way is that given by $\xi^T \eta = 0$. Now, recalling that $\mathscr{V}$ is the quadratic Veronesean of Ω', and that ϕ_1 and ϕ_2 are the Veroneseans on $\mathscr{V}$ corresponding to the generators $y' = 0$ and $x' = 0$ of Ω', we notice that the correlation $\xi^T \eta = 0$ which determines τ_0 corresponds, on Ω', to the *natural correlation*‡ κ' which Ω' induces between $y' = 0$ and $x' = 0$.

More generally, if α_1' and α_2' are any two non-intersecting α'-generators on Ω', then Ω' sets up a natural correlation between α_1' and α_2'; to this there corresponds a uniquely defined correlation between their Veroneseans ϕ_1 and ϕ_2 on $\mathscr{V}$, and hence a unique transformation $\tau(n)$ between the ambient spaces A_1 and A_2 of ϕ_1 and ϕ_2, with ϕ_1 and ϕ_2 as the key base varieties; and the joins of points Q_1 and Q_2 in A_1 and A_2 which correspond by $\tau(n)$ lie on T and represent Clifford reguli in S_{2n-1}.

We can now sum up the results of this section in

THEOREM 3.1. *The T-model of pairs of polar $[n-1]$'s for Ω in S_{2n-1} is uniquely determined by its core variety $\mathscr{V}$, which is the*

---

† Since ϕ_1 and ϕ_2 are Veronesean transforms of two linear spaces, we may regard any correlation correspondence between the linear spaces as defining (by an abuse of language) a *correlation* between ϕ_1 and ϕ_2.

‡ Cf. Ch. 2, §2.

quadratic Veronesean of a non-singular quadric Ω' *of* S'_{2n-1}. *To the generators* α' *and* β' *of* Ω' *there correspond systems of Veroneseans* ϕ *and* ψ *on* $\mathscr{V}$, *and the ambient spaces* A *and* B *of these Veroneseans generate the nuclear primals* $\mathscr{A}$ *and* $\mathscr{B}$ *of* T *whose points correspond in general to the two classes* $\alpha(q)$ *and* $\beta(q)$ *of augmented generators of* Ω *in* S_{2n-1}. *Further,* $\mathscr{V}$ *determines a unique Cremona transformation* $\tau(n)$ *between each non-intersecting pair of the spaces* A *and* B – *of the same kind if* n *is even and of opposite kinds if* n *is odd* – *such that the joins of corresponding points in all such transformations* $\tau(n)$ *lie on* T *and represent Clifford reguli in* S_{2n-1}.

We now illustrate the above results by reference to the cases $n = 2$ and $n = 3$. Consideration of the exceptionally interesting case $n = 4$ will be postponed to Ch. 7, where it can be dealt with alongside other representations of the polar pairs (Π, Π') of S_7, such as arise from two–one projections of the Grassmannian $G(3, 7)$.

The case $n = 2$. The quadrics $\mathscr{Q}(\Pi, \Pi')$ in S_3 are readily identified as those which (a) meet Ω in a (skew) quadrilateral of generators and (b) are apolar to Ω. We find then that T is a $V_4^6[8]$, Segre product of a pair of planes π and π^*, and it lies in a prime of the image space S_9 of the quadrics of S_3. Its core variety $\mathscr{V}$, being the quadratic Veronesean of a quadric Ω' of S'_3, is a Del Pezzo surface $^1F^8$ of the second kind; and it arises on T as image of the product of a conic k in π by a conic k^* in π^*.

The varieties $\mathscr{A}$ and $\mathscr{B}$ on T are the threefolds generated by the two simply infinite families of conic planes A and B of $^1F^8$, and each is therefore a $V_3^6[8]$, Segre product of a conic by a plane. If S_1 and S_2 are any two conics of the first system on $^1F^8$, lying in planes A_1 and A_2, then there is a natural homographic correspondence between S_1 and S_2 determined by their intersections with conics of the second system on $^1F^8$, and this homography is subordinate to a unique collineation $\tau(2)$ between A_1 and A_2 such that the joins of corresponding points lie on V_4^6 and represent Clifford reguli in S_3.

The case $n = 3$. The variety T in this case, as already stated (§ 1, Ex. 4), is the variety V_9^{92}, otherwise studied as the model of quadric complexes of S_3. Its core variety $\mathscr{V}$ is the quadratic

Veronesean of a quadric Ω' of S_5'; the Veronese surfaces on $\mathscr{V}$ corresponding to the generating planes α' and β' of Ω' are such that their ambient [5]'s A and B generate the nuclear† primals $\mathscr{A}$ and $\mathscr{B}$ on T; and if ϕ_0 and ψ_0 are any two of these Veronese surfaces, lying in spaces A_0 and B_0 respectively, then the natural correlation between ϕ_0 and ψ_0 is determined by the Veronese surfaces on $\mathscr{V}$ which meet one of them in a point and the other in a conic, and this correlation is uniquely associated with a Cremona transformation $\tau(3)$ of A_0 into B_0 which carries primes of A_0 into quadrics through ψ_0 and quadrics through ϕ_0 into primes of B_0.

Ex. 1. *The Study representation for $n = 2$.* In connection with the T-model for $n = 2$, as we have described it above, it is of interest to note that an equivalent of this representation is already known in the form of Study's direct mapping‡ of the (unordered) pairs of polar lines (p, p') with respect to a quadric surface Ω in S_3 on the pairs of points (P, P^*) of two planes π and π^*. The construction for this is as follows. Let O be a fixed point not on Ω, and let π and π^* be two fixed planes not through O. Further let the generators of the first system on Ω that meet any general line p be α_1, α_2, and let those of the second system that meet p be β_1, β_2. If the transversals from O to the pairs (α_1, α_2) and (β_1, β_2) meet π and π^* in P and P^* respectively, then we say that to p there corresponds the point-pair (P, P^*); and we observe that the polar line p' of p (and only this line) corresponds in this way to the same point-pair (P, P^*). We have, in fact, a birational mapping of all the (unordered) polar pairs (p, p') on the (ordered) point-pairs (P, P^*) of π and π^*. The connection with our previous mapping will now be apparent, the conics k and k^* of that mapping being identifiable with the conics in which the tangent cone from O to Ω meets the planes π and π^*.

The above representation has the outstanding property that if two polar pairs (p, p') and (q, q') are such that each of p and p'

---

† Semple[21] has shown that V_9^{92} passes sextuply through the 4-fold $\mathscr{V}$ and has no other singularity. The two varieties $\mathscr{A}$ and $\mathscr{B}$ each pass sextuply through $\mathscr{V}$ and they intersect simply in a 7-fold $\mathscr{A}\mathscr{B}$ which passes octuply through $\mathscr{V}$.

‡ Cf. Coxeter[6], p. 146.

meets each of q and q' (i.e. if each of p and p' is perpendicular, in the non-Euclidean sense, to each of q and q'), then their representative point-pairs (P, P^*) and (Q, Q^*) are such that P and Q are conjugate for k, while P^* and Q^* are conjugate for k^*. By this property, any theorem concerning pairs of conjugate points with respect to a conic gives rise to a non-Euclidean theorem concerning perpendicular lines. For example, Hesse's theorem (that a triangle is in perspective with its polar triangle with respect to a given conic) gives rise to a non-Euclidean theorem of which the Euclidean specialization is the well-known Petersen–Morley theorem (that the three common perpendicular transversals of pairs of opposite sides of a skew rectangular hexagon possess a common perpendicular transversal).

4. Fundamental subvarieties of T

The core variety $\mathscr{V}$ of T, as stated in Th. 3.1, is the quadratic Veronesean of a non-singular quadric Ω' of S'_{2n-1}, and the ambient spaces A and B of the Veroneseans ϕ and ψ on $\mathscr{V}$ which correspond to generators α' and β' of Ω' generate the two fundamental primals $\mathscr{A}$ and $\mathscr{B}$ of T. There are, however, lesser fundamental varieties on T generated by the ambient spaces of those Veroneseans on $\mathscr{V}$ that correspond to lines, planes, ..., $[n-2]$'s on Ω'. We denote these by $\mathscr{V}^{(1)}, \mathscr{V}^{(2)}, ..., \mathscr{V}^{(n-2)}$, regarding $\mathscr{V}$ itself as $\mathscr{V}^{(0)}$; and we then have the inclusion diagram

$$\mathscr{V} = \mathscr{V}^{(0)} \subset \mathscr{V}^{(1)} \subset ... \subset \mathscr{V}^{(n-2)} \begin{matrix} \subset \mathscr{A} \\ \subset \mathscr{B} \end{matrix} \subset T.$$

In other words, $\mathscr{V}^{(1)}$ is the locus of conic-planes of $\mathscr{V}$, $\mathscr{V}^{(2)}$ is the locus of [5]'s which meet $\mathscr{V}$ in Veronese surfaces, and so on. We note especially that $\mathscr{V}^{(n-2)}$ is the intersection of $\mathscr{A}$ and $\mathscr{B}$. We now wish to explain the role of these $\mathscr{V}^{(i)}$ in the birational correspondence ρ between the polar pairs (Π, Π') of S_{2n-1} and the points of T.

Since T itself was defined as the variety (locus of specializations) of a certain generic point (cf. (1.2)), and since this point was the coefficient image of the quadric $\mathscr{Q}(\Pi, \Pi')$ with equation (1.1), it follows that T is the coefficient image of an irreducible system of quadrics – to be denoted by $(\mathscr{Q})$ – defined by its generic member $\mathscr{Q}(\Pi, \Pi')$; and we may equally regard ρ as the birational correspondence between polar pairs (Π, Π') and quadrics of the system $(\mathscr{Q})$. Then $(\mathscr{Q})$ will have a set of fundamental subsystems $(\mathscr{Q}^{(0)}), (\mathscr{Q}^{(1)}), \ldots, (\mathscr{Q}^{(n-2)}), (\mathscr{Q}_{\mathscr{A}}^{(n-1)})$ and $(\mathscr{Q}_{\mathscr{B}}^{(n-1)})$ corresponding to the fundamental varieties $\mathscr{V}^{(0)}, \mathscr{V}^{(1)}, \ldots, \mathscr{V}^{(n-2)}, \mathscr{A}$ and $\mathscr{B}$ of T. We now assert:

(i) $(\mathscr{Q}_{\mathscr{A}}^{(n-1)})$ *is the system of quadric cones of* S_{2n-1} *which have an* α-*generator of* Ω (*or a space containing an* α-*generator*) *as vertex; and* $(\mathscr{Q}_{\mathscr{B}}^{(n-1)})$ *is defined similarly;*

(ii) *if* $0 \leqslant h \leqslant n-2$ *and* $k = 2n-2-h$, *then* $(\mathscr{Q}^{(h)})$ *is the system of quadric cones of* S_{2n-1} *which have the polar space* S_k' *of any* S_h *of* Ω *as vertex* (*or whose vertices contain such a space* S_k'); *and, in particular,*

(iii) $(\mathscr{Q}^{(0)})$ *is the system of repeated tangent primes of* Ω.

The significance for the T-representation of the varieties $\mathscr{V}^{(h)}$ ($h = 0, \ldots, n-2$) together with $\mathscr{A}$ and $\mathscr{B}$ can now be stated as follows:

(iv) *For* $h = 0, \ldots, n-2$, $\mathscr{V}^{(h)}$ *is the image variety on* T *of all those specializations of the generic* $\mathscr{Q}$ *that correspond under* ρ *to polar pairs* (Π, Π') *such that* Π *meets* Π' *in a space* S_h *of* Ω. *Similarly* $\mathscr{A}$ *and* $\mathscr{B}$ *are the image varieties on* T *of those specializations of* $\mathscr{Q}$ *that correspond under* ρ *to coincident polar pairs* (α, α) *and* (β, β) *respectively.*

It should be understood, of course, that if Π_0 and Π_0' meet in an S_h of Ω, and if $\overline{P}$ is the image on T of a $\overline{\mathscr{Q}}$ which is one of those arising under ρ from (Π_0, Π_0'), then (iv) only tells us that $\overline{P}$ lies in a certain generator of $\mathscr{V}^{(h)}$. Its position in this generator will depend on the 'approach path' from the generic (Π, Π') to (Π_0, Π_0'). Also the same point $\overline{P}$ may correspond under ρ to other polar pairs (Π_1, Π_1') such that Π_1 meets Π_1' in S_h.

We shall now indicate in outline how the above statements (i), ..., (iv) can be verified, and we begin by proving

PROPOSITION 4.1. *If* γ *is any* $[n-1]$-*generator of* Ω, *then every quadric cone with vertex* γ (*or with vertex containing* γ) *is a specialization of the generic* $\mathscr{Q}(\Pi, \Pi')$.

Proof. By equation (2.1) of §2, the equation of the generic $\mathcal{Q}(\Pi, \Pi')$ can be written in the form

$$x^T(C - SC^{-1}S)\,x + 2x^T SC^{-1}y + y^T C^{-1}y = 0, \qquad (4.1)$$

where the matrices C and S are symmetric and skew-symmetric respectively, with independent indeterminates as elements. Let γ be the generator $y = S_0 x$, and let $C + S$ be specialized to $tD^{-1} + S_0$, where D is a symmetric matrix of indeterminates and t is a further indeterminate. From (4.1), the resulting specialization $\bar{\mathcal{Q}}$ of $\mathcal{Q}(\Pi, \Pi')$ has equation

$$t^2 x^T D^{-1} x + \{-x^T S_0 D S_0 x + 2x^T S_0 D y + y^T D y\} = 0;$$

and when $t \to 0$, we get the further specialization $\mathcal{Q}^*$ given by

$$-x^T S_0 D S_0 x + 2x^T S_0 D y + y^T D y = 0,$$

which can be written in the form

$$(y - S_0 x)^T\, D(y - S_0 x) = 0. \qquad (4.2)$$

Thus (4.2) is a specialization of $\mathcal{Q}(\Pi, \Pi')$, and since it represents the generic cone with vertex γ, the proof of the Proposition is complete.

Since we proved earlier that $\mathscr{A}$ (resp. $\mathscr{B}$) is the fundamental variety of T which corresponds by ρ to all coincidence polar pairs (α, α) (resp. (β, β)), Prop. 4.1 is equivalent to statement (i).

We now prove

PROPOSITION 4.2. *If* (Π, Π') *is a generic polar pair and* (Π_0, Π_0') *is a polar pair such that* Π_0 *meets* Π_0' *in a space* S_h $(0 \leqslant h \leqslant n-1)$ *on* Ω, *then as* $(\Pi, \Pi') \to (\Pi_0, \Pi_0')$ *in any manner, the quadric* $\mathcal{Q}(\Pi, \Pi') \to$ *a quadric cone* $\bar{\mathcal{Q}}$ *with the join of* Π_0 *to* Π_0' *(or a space containing this join) as vertex.*

Proof. Since S_h lies on Ω, the join of Π_0 to Π_0' is the polar space S_k' of S_h, where $k = 2n - 2 - h$.

Let the sections of Ω by Π and Π' be q and q', and let K_1 and K_2 be the cones $\Pi(q')$ and $\Pi'(q)$. Then $\mathcal{Q}(\Pi, \Pi') \equiv \mathcal{Q}$ is the harmonic conjugate of Ω with respect to K_1 and K_2 in the pencil they determine. Now let $(\Pi, \Pi') \to (\Pi_0, \Pi_0')$ in any manner, and let the corresponding specializations of K_1 and K_2 be $\bar{K}_1$ and $\bar{K}_2$.

Then since Π is a double locus on K_1, Π_0 will be double on $\bar{K}_1$; and similarly Π'_0 will be a double locus on $\bar{K}_2$. Thus, in particular, S_h will be double on each of $\bar{K}_1$ and $\bar{K}_2$. Now, since K_1, K_2 and Ω are linearly dependent, so in the limit must $\bar{K}_1$, $\bar{K}_2$ and Ω be linearly dependent. If $\bar{K}_1$ and $\bar{K}_2$ were distinct, then the linear dependence would require that S_h, being double for both, would also be double for Ω – an impossibility since Ω is non-singular. Hence $\bar{K}_1$ and $\bar{K}_2$ coincide, and we write $\bar{K}_1 \equiv \bar{K}_2 = \bar{K}$.

Further, since $\mathscr{Q}$ and Ω separate K_1 and K_2 harmonically, it follows now that the limit $\bar{\mathscr{Q}}$ of $\mathscr{Q}$, since it must be the harmonic conjugate of Ω with respect to $\bar{K}_1$ and $\bar{K}_2$, is $\bar{K}$. Also, since $\bar{K}$, regarded as $\bar{K}_1$, will have Π_0 as double locus, and, regarded as $\bar{K}_2$, will have Π'_0 as double locus, it follows that $\bar{K}$ (i.e. $\bar{\mathscr{Q}}$) has the join of Π_0 and Π'_0 as double locus.

Hence, as $(\Pi, \Pi') \to (\Pi_0, \Pi'_0)$, the resulting specialization $\bar{\mathscr{Q}}$ of $\mathscr{Q}(\Pi, \Pi')$ is a quadric with the join S'_k of Π_0 and Π'_0 (or a space containing this join) as vertex. This completes the proof.

To carry the argument further, we recall from § 3 (cf. equations (3.3) and (3.4) and the remarks there following) that the points of the core variety $\mathscr{V}$ of T are the coefficient images of the repeated tangent primes of Ω, the latter being extreme specializations of the generic $\mathscr{Q}(\Pi, \Pi')$. Thus to any S_h on Ω, i.e. to the aggregate of repeated tangent primes of Ω at points of S_h, there corresponds on $\mathscr{V}$ an h-dimensional quadric Veronesean† V_h whose ambient space is a $[\frac{1}{2}h(h+3)]$. Now to any generator α of Ω that passes through S_h there corresponds by ρ a space A; and this space A must clearly contain V_h and hence also its ambient $[\frac{1}{2}h(h+3)]$. It readily appears, then, that *this $[\frac{1}{2}h(h+3)]$ is the intersection of all the spaces A which correspond to α-generators of Ω through S_h.*

Now consider again the quadric $\bar{\mathscr{Q}} = \bar{K}$ which arose in the proof of Prop. 4.2 as a specialization of the generic $\mathscr{Q}$ arising from a polar pair (Π_0, Π'_0) such that Π_0 meets Π'_0 in S_h. Since this $\bar{\mathscr{Q}}$ had vertex S'_k – the polar of S_h – it follows that every α-generator through S_h will lie in S'_k; and hence $\bar{\mathscr{Q}}$ can be regarded (for each

---

† The quadric Ω' (of which $\mathscr{V}$ is the quadratic Veronesean) is, as we have seen in §3, in dual correspondence with Ω, the tangent primes of Ω being in natural correspondence with the points of Ω'. Hence the tangent primes of Ω at points of S_h correspond on Ω' to the points of an $[h]$, and V_h is the quadric Veronesean of this $[h]$.

such α) as a cone with that α-generator as vertex. Hence *the image point of $\overline{\mathcal{Q}}$ on T must lie in the intersection of all the spaces A that correspond to α-generators through S_h.*

Combining now the last two observations, and noting also that the freedom of quadric cones with S'_k as vertex is $\frac{1}{2}h(h+3)$, we deduce that all the $\overline{\mathcal{Q}}$ arising from polar pairs which intersect in S_h are represented on T by the points of the ambient $[\frac{1}{2}h(h+3)]$ of the quadric Veronesean V_h on $\mathscr{V}$. Thus, letting S_h vary on Ω, it follows that the image variety on T of all specializations of the generic $\mathcal{Q}$ that arise from polar pairs which intersect in an S_h is the fundamental variety $\mathscr{V}^{(h)}$ of T. This completes, therefore, our verification of the statements (i), ..., (iv) which we made at the beginning of this section.

5. Self-collineations of a linear system $\mathscr{F}$

In this section we take up again from § 2 the discussion of an r-dimensional linear system $\mathscr{F}$ of Clifford parallels of S_{2n-1} and the mapping there described of the polar pairs (Π, Π') of $\mathscr{F}$ on the points of a linear space $L(\mathscr{F})$ on the model T. We then use this mapping to discuss the group Γ of permutations on the spaces of $\mathscr{F}$ that are induced by self-collineations of S_{2n-1} which leave Ω and $\mathscr{F}$ invariant. From this discussion there will emerge certain structural homogeneity properties of those linear systems $\mathscr{F}$ which we have termed 'non-singular', emphasizing their relative simplicity and importance.

We recall, then, the previous mapping of the polar pairs (Π, Π') of $\mathscr{F}$ on the points of $L(\mathscr{F})$ as follows. We took the canonical equation of $\mathscr{F}$, subject to the conditions (3.1) of Ch. 4, in the form

$$y = (\alpha I + S(\lambda))\,x,$$

where

$$S(\lambda) = \sum_{i=1}^{r-1} \lambda_i S_i$$

and we saw then that a coordinate system $\xi_0, ..., \xi_r$ in $L(\mathscr{F})$ could be chosen in such a way that the polar pair given by $y = (\pm \alpha I + S(\lambda))\,x$ was represented in $L(\mathscr{F})$ by the point

$$(\xi_0, ..., \xi_r) = (1, \lambda_1, \lambda_2, ..., \lambda_{r-1}, \alpha^2 + \lambda_1^2 + ... + \lambda_{k-1}^2).$$

This implies, in particular, that the spaces of $\mathscr{F}$ which meet the reference space Y can only be represented in $L(\mathscr{F})$ by points lying in the 'prime at infinity' $\xi_0 = 0$ in this space. Further, we saw that the augmented generators of $\mathscr{F}$, based on generators of Ω with equations of the form $y = S(\lambda)\,x$, were represented in $L(\mathscr{F})$ by points of the quadric $\mathscr{N}(\mathscr{F})$ with equation

$$\xi_0\xi_r = \xi_1^2 + \ldots + \xi_{k-1}^2.$$

We note here also that of all the augmented generators based on a generator $y = S(\lambda)\,x$ contained in $\mathscr{F}$ only one – the *augmented generator properly belonging to $\mathscr{F}$* (cf. Ch. 4, remark following Cor. 3.1.1) – is represented in $L(\mathscr{F})$ by a point, and this point lies on $\mathscr{N}(\mathscr{F})$.

Now consider the permutations on the spaces of $\mathscr{F}$ that arise from self-collineations of S_{2n-1} which leave Ω and $\mathscr{F}$ invariant. We begin with the remark:

Any self-collineation of S_{2n-1} which leaves Ω invariant – a (complex) non-Euclidean 'congruence transformation' – induces a self-collineation of T.

For, in the first place, any self-collineation ϖ of S_{2n-1} induces a self-collineation ϖ' of the space S_N of quadric primals of S_{2n-1}; and if ϖ leaves Ω invariant, so that it carries polar pairs always into polar pairs, then ϖ' must carry T (a variety in S_N) into itself.

From this it follows that if ϖ also leaves $\mathscr{F}$ invariant, then ϖ' must leave $L(\mathscr{F})$ invariant, i.e. it induces a self-collineation of $L(\mathscr{F})$; and further, since ϖ permutes the (augmented) generators in $\mathscr{F}$ among themselves, it follows that ϖ' must transform the quadric $\mathscr{N}(\mathscr{F})$ into itself. This proves

PROPOSITION 5.1. *If $\mathscr{F}$ is any linear system of parallels in S_{2n-1}, then every self-collineation of S_{2n-1} which leaves Ω and $\mathscr{F}$ invariant induces a self-collineation of $L(\mathscr{F})$ which leaves the quadric $\mathscr{N}(\mathscr{F})$ invariant.*

It is important to notice that the self-collineation of $L(\mathscr{F})$ may be the identity, even when the self-collineation of S_{2n-1} from which it derives is not. Thus there is a difference between the group of self-collineations of S_{2n-1} which leave Ω and $\mathscr{F}$ invariant and the group of permutations of the spaces of $\mathscr{F}$ induced by such self-collineations.

Now consider any chordal polar pair (Π_0, Π_0') in S_{2n-1}, and let $\varpi(\Pi_0, \Pi_0')$ denote the biaxial harmonic homology in S_{2n-1} with Π_0 and Π_0' as axial spaces – a type of collineation which we may refer to as a *skew reflection* with respect to Ω. Plainly $\varpi(\Pi_0, \Pi_0')$ leaves Ω invariant and carries polar pairs into polar pairs. Suppose then that (Π_0, Π_0') belongs to $\mathscr{F}$. Then if (Π, Π') is a generic polar pair of $\mathscr{F}$, there exists a Clifford regulus R containing (Π_0, Π_0') and (Π, Π') and wholly contained in $\mathscr{F}$. Hence $\varpi(\Pi_0, \Pi_0')$ transforms R into itself, permuting the polar pairs of R among themselves (cf. Ch. 3, §4, Ex. 1). Hence:

If (Π_0, Π_0') is any polar chordal pair of $\mathscr{F}$, then the skew reflection with Π_0, Π_0' as axial spaces leaves Ω and $\mathscr{F}$ invariant, permuting polar pairs of $\mathscr{F}$ among themselves.

We now prove

PROPOSITION 5.2. *If (Π_0, Π_0') is any chordal polar pair of a linear system $\mathscr{F}$, then the skew reflection $\varpi(\Pi_0, \Pi_0')$ induces in $L(\mathscr{F})$ a central harmonic homology $\varpi'(P_0)$ whose centre P_0 is the image point of (Π_0, Π_0') and whose axial prime is the polar of P_0 with respect to the quadric $\mathscr{N}(\mathscr{F})$.*

Plainly $\varpi'(P_0)$ leaves $\mathscr{N}(\mathscr{F})$ invariant, and we shall call it a *central reflection* in $L(\mathscr{F})$ with respect to $\mathscr{N}(\mathscr{F})$.

Proof. Let $\varpi(\Pi_0, \Pi_0')$ carry a generic polar pair (Π_1, Π_1') of $\mathscr{F}$ into (Π_2, Π_2'). Then, by what we have said above, all the six spaces concerned belong to a Clifford regulus R contained in $\mathscr{F}$; and we can choose a coordinate system (Ch. 3, Th. 4.2) in which Ω has equation $x^T y = 0$ and R is given by $y = \theta x$. Let Π_0, Π_0' be given in R by $\theta = \pm\alpha$. Then (cf. Ch. 3, §4, Ex. 1) $\varpi(\Pi_0, \Pi_0')$ carries $y = \theta x$ into $y = (\alpha^2/\theta)\,x$; so that, if Π_1, Π_1' are given by $y = \pm\theta_1 x$, then Π_2, Π_2' are given by $y = \pm(\alpha^2/\theta_1)\,x$. Now, on the T-model, the pairs of polar spaces $y = \pm\theta x$ are represented by the points of a line on which $\phi = \theta^2$ is a projective parameter. The pair (Π_0, Π_0') is represented on this line by the point P_0 for which $\phi = \alpha^2$, the pair (Π_1, Π_1') by the point P_1 given by $\phi = \theta_1^2$; and the pair (Π_2, Π_2') by the point P_2 for which $\phi = \alpha^4/\theta_1^2$. Now the correspondence between P_1 and P_2, given by $\phi \to \alpha^4/\phi$, is involutory with united points given by $\phi = \pm \alpha^2$. One of these united points $(\phi = \alpha^2)$ is P_0, and the other $(\phi = -\alpha^2)$ is the harmonic conjugate of P_0 with respect to the points given by $\phi = 0$ and $\phi = \infty$, these latter being evidently the two points of $\mathcal{N}(\mathcal{F})$ which represent the terminal spaces of R (given by $\theta = 0$ and $\theta = \infty$). In other words, the harmonic conjugate of P_0 with respect to P_1 and P_2 lies on the polar of P_0 with respect to $\mathcal{N}(\mathcal{F})$. Thus the transformation carrying P_1 into P_2 is the central reflection (relative to $\mathcal{N}(\mathcal{F})$) with P_0 as vertex. This completes the proof.

For the special case in which the linear system $\mathcal{F}$ is non-singular, the result just proved has certain important implications. In this case, as already remarked, $\mathcal{N}(\mathcal{F})$ is a non-singular quadric in $L(\mathcal{F})$; and hence, by a known theorem,† the central reflections in $L(\mathcal{F})$ with respect to $\mathcal{N}(\mathcal{F})$ generate the whole group Γ' of self-collineations of $L(\mathcal{F})$ which leave $\mathcal{N}(\mathcal{F})$ invariant. Hence, when $\mathcal{F}$ is non-singular, every self-collineation of $\mathcal{N}(\mathcal{F})$ corresponds to a permutation of the spaces of $\mathcal{F}$ which is generated by a finite sequence of the skew reflections $\varpi(\Pi, \Pi')$ of S_{2n-1} such that (Π, Π') is a chordal polar pair of $\mathcal{F}$. Also, conversely, by Prop. 5.1, every permutation of the spaces of $\mathcal{F}$ which is generated by a self-collineation of S_{2n-1} leaving Ω and $\mathcal{F}$ invariant corresponds to a self-collineation of $L(\mathcal{F})$ belonging to the group Γ'. Hence:

COROLLARY 5.2.1. *When $\mathcal{F}$ is non-singular, the group Γ of permutations of the spaces of $\mathcal{F}$ arising from self-collineations of S_{2n-1} which leave Ω and $\mathcal{F}$ invariant is isomorphic with the group Γ' of self-collineations of $L(\mathcal{F})$ which leave $\mathcal{N}(\mathcal{F})$ invariant.*

But now further (still supposing $\mathcal{F}$ to be non-singular), we remark that the group Γ' is transitive at two levels on the points of $L(\mathcal{F})$, distinguishing only between points of $L(\mathcal{F})$ which do or do not lie on $\mathcal{N}(\mathcal{F})$. Thus in particular, if P is any point in the 'prime at infinity' $\xi_0 = 0$ of our previous construction, there exists a self-collineation of Γ' which carries P into a 'finite' point

† Cf. Burau[4], p. 234.

P', image either of a chordal polar pair or of an augmented generator of $\mathscr{F}$. It follows therefore, since either of these latter is carried by any permutation of Γ into another of the same kind, that every point such as P (like every 'finite' point of $L(\mathscr{F})$) corresponds either to a chordal polar pair or to an augmented generator of $\mathscr{F}$. This gives

COROLLARY 5.2.2. *If $\mathscr{F}$ is non-singular, then all the points of $L(\mathscr{F})$, including those which lie in the 'prime at infinity' $\xi_0 = 0$, represent either chordal pairs or augmented generators of $\mathscr{F}$, and $\mathscr{F}$ consists of such spaces and no others. The permutation group Γ of $\mathscr{F}$ is transitive on each of the two types of spaces of $\mathscr{F}$.*

The group Γ', we now note, is also transitive on the lines of $L(\mathscr{F})$ which meet $\mathscr{N}(\mathscr{F})$ in two distinct points. It follows that *every* such line represents a Clifford regulus within $\mathscr{F}$ and, in particular, that every space of $\mathscr{F}$ belongs to some Clifford regulus within $\mathscr{F}$. Hence, by Ch. 4, Cor. 4.1.1, we deduce that a non-singular system $\mathscr{F}$ cannot contain any singular space. This (in conjunction with the remark concerning singular spaces after Cor. 2.2.1 of this Chapter) proves

COROLLARY 5.2.3. *A linear system $\mathscr{F}$ is non-singular if and only if it contains no singular spaces.*

Finally, for a non-singular system $\mathscr{F}$, we note that if Σ is a terminal (augmented) generator of any Clifford regulus within $\mathscr{F}$, represented in $L(\mathscr{F})$ by a point P of $\mathscr{N}(\mathscr{F})$, then the Σ-affine part of $\mathscr{F}$ (as defined in Ch. 4, §4) is represented by the points of $L(\mathscr{F})$ that do not lie in the tangent prime to $\mathscr{N}(\mathscr{F})$ at P. Now if $\mathscr{F}$ has freedom $r > 1$, and if P_1 and P_2 are two arbitrary points of $L(\mathscr{F})$, we can always find a point P of $\mathscr{N}(\mathscr{F})$ such that the tangent prime to $\mathscr{N}(\mathscr{F})$ at P does not pass through P_1 or P_2. This implies then that, given two arbitrary spaces Π_1, Π_2 of $\mathscr{F}$, it is always possible to find a Σ (terminal in some Clifford regulus of $\mathscr{F}$) such that Π_1 and Π_2 belong to the Σ-affine part of $\mathscr{F}$. Hence, by Prop. 4.2 of Ch. 4, we deduce

COROLLARY 5.2.4. *If two distinct spaces of a non-singular linear system intersect, then their intersection lies on Ω.*

Corollaries 5.2.3 and 5.2.4, it will be noted, give final confirmation to two results which were mentioned without proof in Ch. 4, § 4.

If $\mathscr{F}$ is a singular linear system, then $\mathscr{N}(\mathscr{F})$ is a cone; the central reflections with respect to $\mathscr{N}(\mathscr{F})$ no longer generate the whole group Γ' of self-collineations of $\mathscr{N}(\mathscr{F})$, and it need no longer be true that all the spaces of $\mathscr{F}$ are either chordal or augmented generators. In other words, the prime $\xi_0 = 0$ may contain points – possibly fundamental for the T-representation – which represent spaces of $\mathscr{F}$ that are neither chordal nor augmented generators; and in fact we have already noted the existence of singular systems which possess spaces of this kind (Ch. 4, § 4, Ex. 4). On the other hand if a singular system $\mathscr{F}$ is such that it is contained in a more ample non-singular linear system $\mathscr{F}'$, then the first conclusion of Cor. 5.2.2, being valid for $\mathscr{F}'$, is necessarily also valid for $\mathscr{F}$.

The relatively simple structure of non-singular linear systems, as set out in Cor. 5.2.2, is the main justification for the emphasis we have placed on them throughout this book.

E x. 1. If (Π_0, Π_0') is the polar pair given by $y = (\pm \alpha I + S(\lambda))\, x$, where $S(\lambda) = \sum\limits_{i=1}^{r-1} \lambda_i S_i$, and if Π is the $[n-1]$ given by

$$y = (\beta I + S(\mu))\, x,$$

show that the transform Π_1 of Π by the skew-reflection $\varpi(\Pi_0, \Pi_0')$ has equation

$$y = \phi^{-1}\{\beta I + (\phi + 1)\, S(\lambda) - S(\mu)\}\, x$$

where

$$\phi = \frac{1}{\alpha^2}\Big\{\beta^2 + \sum\limits_{i=1}^{k-1} (\mu_i - \lambda_i)^2\Big\}.$$

CHAPTER 7

HALF-GRASSMANNIANS

In this final chapter we discuss a representational method which is different from that described in Ch. 6, employing two new models of the polar pairs (Π, Π') of S_{2n-1}; and these new models will be called half-Grassmannians because they are, in effect, $(2, 1)$-projections of the Grassmannian $G(n-1, 2n-1)$ of all $[n-1]$'s of S_{2n-1}. After a rapid survey of the general theory of this method and brief discussions of the cases $n = 2$ and $n = 3$, we treat in some detail the case $n = 4$ which is in many ways unique. This leads us directly to a simple construction and interpretation of the so-called Study triality correspondence between the points, α-solids and β-solids of a quadric in S_7. Only in this case $(n = 4)$ does it happen that each of the two half-Grassmannians concerned is projectively equivalent to the corresponding T-model; and the birational (but non-linear) correspondences between pairs of the three models that result from the three different representations give rise to an extended form of the Study triality.

1. Grassmannian representation

We denote by G the Grassmannian $G(n-1, 2n-1)$ of all $[n-1]$'s of S_{2n-1}, noting that G is of dimension n^2 and that its ambient space is S_N, where $N = \binom{2n}{n} - 1$.

Further, we denote by G_α and G_β the subvarieties of G that map the systems of α-generators and β-generators of Ω. The properties of a variety such as G_α or G_β, which represents the generators of one system of a quadric in S_{2n-1}, are well known.[†] Thus, for example, G_α has a minimum model M_α of which it is precisely the quadratic Veronesean. In particular for $n = 2$, M_α is a line and G_α is a conic; for $n = 3$, M_α is a solid and G_α is a

† See, for example, Burau[4], Ch. 9 and Heymans[11, 12].

Veronesean $V_3^8[9]$; while for $n = 4$, M_α is a quadric Q_6^2 of S_7 and G_α is its quadratic Veronesean[†] $V_6^{128}[34]$. For the moment we note only that G_α and G_β are each of dimension $\frac{1}{2}n(n-1)$; that their ambient spaces, which we shall denote by L and M respectively, are each of dimension $\dfrac{1}{2}\dbinom{2n}{n} - 1 = \frac{1}{2}(N-1)$; and that these ambient spaces, which meet G only in G_α and G_β respectively, span the ambient space S_N of G.

We now prove

PROPOSITION 1.1. *If P and P' are the image points of polar $[n-1]$'s Π and Π' with respect to Ω, then P and P' correspond in the harmonic homology of S_N with L and M as axial spaces.*

Proof. Let the equation of Ω, in a suitably chosen coordinate system of S_{2n-1}, be

$$z_0^2 + z_1^2 + \ldots + z_{2n-1}^2 = 0.$$

If Π is a generic $[n-1]$ of S_{2n-1}, join of the n points with co-ordinate vectors ξ_i $(i = 0, \ldots, n-1)$, then its polar space Π' is the intersection of the n primes with the same coordinate vectors ξ_i. This means, by the well-known relations between ordinary and dual Grassmann coordinates,[‡] that if $p_0, \ldots, p_N$ are a suitably arranged set of coordinates for Π, then the coordinates $p_0', \ldots, p_N'$ of Π' are equal to those of Π but interchanged by pairs, i.e.

$$p_\alpha' = p_\beta, \quad p_\beta' = p_\alpha,$$

where the $\frac{1}{2}(N+1)$ pairs (α, β) together make up the set $(0, 1, \ldots, N)$. Plainly then the image points P, P' of Π, Π' correspond in the harmonic homology of S_N with axial spaces L and M given respectively by

$$p_\alpha + p_\beta = 0 \quad \text{(all pairs } (\alpha, \beta))$$

and $\qquad\qquad p_\alpha - p_\beta = 0 \quad \text{(all pairs } (\alpha, \beta)).$

Also L and M contain, evidently, the image varieties G_α and G_β, respectively, of the two systems of generators (self-polar $[n-1]$'s) of Ω. This proves the Proposition.

[†] For the proof in this case $(n = 4)$, see Ch. 5, Prop. 4.1.
[‡] See Hodge and Pedoe[13], Vol. 1, p. 294.

By a straightforward reversal of the above argument we have

COROLLARY 1.1.1. *If two points P, P' of G correspond in the harmonic homology of S_N with axial spaces L and M, then they represent spaces Π and Π' which are polars with respect to Ω.*

We now denote by σ the involution on G, with coincidence loci G_α and G_β, whose point-pairs represent pairs of polar spaces for Ω. Then two distinct points P, P' of G correspond in σ if and only if their join meets L and M in points U, V such that

$$\{P, P'; U, V\} = -1.$$

Further, if P is any point of G not lying on G_α or G_β, then the unique transversal line of L and M through P contains the point P' of G which corresponds in σ to P. If a line meets G in more than two points, then it lies entirely on G – as follows from the well-known result (Ch. 2, §5) that any Grassmannian is an intersection of quadrics – and it then represents a pencil of $[n-1]$'s of S_{2n-1} (lying in an $[n]$ and passing through an $[n-2]$). Thus we may assert

COROLLARY 1.1.2. *The pairs (P, P') of the involution σ are in birational (though not unexceptional) correspondence with the chords of G which meet L and M.*

The exceptional elements in this correspondence arise from

(a) the points of G_α and G_β, since P coincides with P' at any such point so that the chord PP' is indeterminate, and

(b) the lines which lie on G and join a point of G_α to a point of G_β.

As regards (b), if p is one of the lines in question, then it represents a pencil of $[n-1]$'s, defined by a pair of generators α, β of Ω which meet in an $[n-2]$ and lie in an $[n]$; and then p contains an infinity of σ-pairs representing pairs of $[n-1]$'s of this pencil, polars for Ω, which separate α and β harmonically.

We now consider, as arising from Cor. 1.1.2, the correspondence between the σ-chords PP' – a system of dimension n^2 – and the points in which these chords meet one of the axial spaces, say L, which is of dimension $\frac{1}{2}\binom{2n}{n} - 1$. Plainly, if $n = 2$ so that L is a plane, there must be infinitely many σ-chords through each point of L. We now assert however that if $n \geqslant 3$, so that

$$\frac{1}{2}\binom{2n}{n} - 1 \geqslant n^2,$$

then the σ-chords are in birational correspondence with the points in which they meet L. For any contrary hypothesis† would imply that there exist couples of σ-pairs, say (P, P') and (P_1, P_1'), representing pairs of skew polar $[n-1]$'s (Π, Π') and (Π_1, Π_1'), such that the σ-chords PP' and P_1P_1' meet on L; this again would imply that the four spaces Π, Π', Π_1, Π_1', being linearly dependent, have the property that every $[n-1]$ meeting three of them would also meet the fourth; and this, finally (cf. Ex. 1 below), can be shown to be impossible for $n > 2$ and under the conditions stated (Π skew to Π' and Π_1 skew to Π_1'). From this there follows

COROLLARY 1.1.3. *If $n \geqslant 3$, then the pairs (P, P') of the involution σ on G are in birational correspondence with the points of each of the two n^2-dimensional varieties which their joins PP' cut on L and M respectively.*

The two n^2-dimensional varieties mentioned in the above Corollary will be called *half-Grassmannians*.

By combining Cor. 1.1.2 and Cor. 1.1.3 we now obtain

COROLLARY 1.1.4. *If $n \geqslant 3$, then G projects doubly (i.e. by a $(2, 1)$-projection) from either of the axial spaces L or M into a half-Grassmannian variety Δ or Γ in the other. Each of these half-Grassmannians Γ, Δ becomes thereby a birational model of the polar pairs (Π, Π') of S_{2n-1} with respect to Ω.*

We now consider those curves on G – to be called C-curves – which represent Clifford reguli of S_{2n-1}; and we prove

PROPOSITION 1.2. *The spaces of a regulus R of $[n-1]$'s of S_{2n-1} are represented on G by the points of a rational normal curve*

---

† The possible contrary hypotheses are that the projection of G from M onto a variety Γ in L is (*a*) dimension-lowering or (*b*) dimension-preserving but $(2N, 1)$ with $N > 1$; but both of these hypotheses lead easily to the implication as stated.

of order n; and R is a Clifford regulus if and only if this curve is compounded of the involution σ.

Proof. Without loss of generality we may suppose that the equation of R, in a suitably chosen coordinate system for S_{2n-1}, is $y = \lambda x$ (cf. Ch. 2, § 3). A simple calculation then shows that (apart from sign) the Grassmann coordinates of $y = \lambda x$ are just repetitions of $0, 1, \lambda, \ldots, \lambda^n$, and this proves the first assertion. The second follows at once from Cor. 1.1.1 using the autopolar property of a Clifford regulus.

If R is a Clifford regulus, then to the two terminal generators of R there will correspond two *terminal points* on the corresponding C-curve. Also, if n is even, these terminal points will either both lie on G_α or both on G_β; while, if n is odd, one of them will lie on G_α and the other on G_β. Hence:

COROLLARY 1.2.1. *If n is even, then the C-curves of one family on G meet G_α in two points and project doubly from L into curves of order $\frac{1}{2}(n-2)$ in M (not in general meeting G_β); while those of the second family (not meeting G_α) project doubly from L into curves of order $\frac{1}{2}n$ in M, each meeting G_β in two points. If n is odd, then the C-curves on G meet each of G_α and G_β in one point and project doubly from L into curves of order $\frac{1}{2}(n-1)$ in M, each meeting G_β in one point.*

We now note briefly, for reference, some general properties of the Grassmannian representation for the case of arbitrary $n \geqslant 3$. First, as regards the C-curves, and taking account of results already obtained in Ch. 3, Cor. 4.1.1, we note the following:

(i) The total freedom of C-curves on G, whether of two families for n even, or of one family for n odd, is $\frac{1}{2}(n-1)(3n+2)$.

(ii) The freedom of C-curves (of either family if n is even) through a general σ-pair (P, P') is $\frac{1}{2}n(n-1)$.

(iii) The freedom of C-curves through a general point of G_α or G_β is $n^2 - 1$.

(iv) The freedom of C-curves with given terminal points is $\frac{1}{2}(n-1)(n+2)$.

Next, as regards the half-Grassmannians Γ, Δ, in L, M respectively, we note:

(v) If P is any point of G_α, representing a generator α of Ω, then the generators β which each meet α in an $[n-2]$ are mapped on G_β by the points of a variety ψ_P which is the quadric Veronesean of an $[n-1]$.

(vi) In the (double) projection of G from L into Δ, the neighbourhood on G of any point P of G_α is projected into the ambient space – to be denoted by K_P – of the associated variety ψ_P of G_β; and the whole neighbourhood of G_α on G is projected accordingly into a primal $((n^2-1)$-dimensional subvariety) of Δ – to be denoted by $\mathscr{K}$ – which is generated by the spaces K_P.

(vii) If P is any point of G_α, then the $[n]$'s that lie on G and pass through P fall into two systems; an $[n]$ of the first system represents the $[n-1]$'s of S_{2n-1} that lie in a fixed $[n]$ through the generator α corresponding to P, and an $[n]$ of the second system represents the $[n-1]$'s that pass through a fixed $[n-2]$ contained in α. Every line joining P to a point Q of ψ_P is the intersection of a unique pair of $[n]$'s of the two systems, and the two $[n]$'s of this pair project from L into the same tangent $[n-1]$ to ψ_P at Q.

We may note, finally, that if any C-curve passes through a point P of G_α, then its projection from L meets the space K_P which corresponds on Δ to the neighbourhood of P on G. Under this projection, in fact, the points of the space K_P represent (in general) the augmented generators based on the generator α represented by P.

Ex. 1. If $n \geqslant 3$ and if Π, Π', Π_1, Π_1' are four distinct $[n-1]$'s of S_{2n-1} such that Π is skew to Π' and Π_1 is skew to Π_1', prove that the four spaces cannot be such that every $[n-1]$ which meets three of them also meets the fourth.

[A proof along the following lines is suggested. First observe that, if Π_1' meets every $[n-1]$ which meets Π, Π' and Π_1, then it meets every *line* which meets these three spaces. Distinguish then between the cases (a) when all four spaces are mutually skew, and (b) when, say, Π meets Π_1 in a space S_h of dimension h. In case (a), by our first observation, the four spaces belong to a regulus; and this, by Prop. 1.2, is represented on G by a rational normal $^0C^n$ which contains no set of four coplanar points if

$n > 2$. In case (b), if $h \geqslant 1$, then the same observation tells us that Π_1' must meet every line joining a point of S_h to a point of Π'; and it must therefore contain S_h, in contradiction of the assumption that Π_1 is skew to Π_1'; and if, on the other hand, $h = 0$, then either Π_1' contains S_0 and the same argument applies; or Π_1' lies in the $[n]$ joining S_0 to Π', in which case Π_1' meets Π' in an $[n-2]$, and the previous argument applies to this intersecting pair.]

2. The cases $n = 2$ and $n = 3$

The case $n = 2$. This case requires only brief mention since it has already been dealt with in some detail (Ch. 4, § 1, Ex. 1).

We note, then, that G is here a quadric of S_5; and G_α, G_β are the conics in which G is met by a pair of skew polar planes L, M. Also σ is the involution of point-pairs of G whose joins meet L and M.

In this case, therefore, there is a birational correspondence between the point-pairs (P, P') of σ and the pairs of points (U, V) in which their joins PP' meet L and M respectively; and this is equivalent to a birational correspondence† between pairs of polar lines with respect to Ω and the point-pairs (U, V) of L and M. The exceptional elements arise from pairs (U, V) for which U lies on G_α or V on G_β, and also from pairs for which UV lies on G (i.e. U and V lie respectively on G_α and G_β).

The two types of Clifford reguli in S_3 (Ch. 4, § 1, Ex. 1) are represented on G by conics – sections of G by planes which meet one of L, M in a line and the other in a point. The two types of (maximal) linear ∞^2-systems in S_3 are represented by quadric surfaces – sections of G by solids through L and solids through M.

As already noted, the case $n = 2$ is exceptional; for only in this case does it happen that infinitely many of the chords PP' pass through any general point of L or M.

The case $n = 3$. Here Ω is a quadric V_4^2 of S_5; G is a variety V_9^{42} in S_{19}; the two systems of planes on Ω are mapped on G by varieties G_α and G_β which are quadric Veroneseans V_3^8 lying in skew 9-dimensional spaces L and M of S_{19}; and the involution σ on G is generated by the chords of G that meet L and M. In

† This is again, effectively, the Study correspondence, cf. Ch. 6, §3, Ex. 1.

this case (cf. Cor. 1.1.3 and Cor. 1.1.4), only one such chord passes through a general point of L or M, i.e. *L and M are themselves the half-Grassmannians Γ and Δ – double projections of G from M and L respectively.* Hence:

PROPOSITION 2.1. *When $n = 3$, G projects doubly from L onto M and from M onto L, each of the spaces L, M being therefore a birational model of the pairs of polar planes with respect to the quadric Ω in S_5.*

The $(2, 1)$-projection of G from L onto M has exceptional elements. In particular, the tangent [9] to G at a point P of G_α, since it meets L in the tangent solid to G_α at P, projects into a 5-dimensional space K_P of M; and all such spaces K_P generate a primal $\mathcal{K}$ of M representing the whole neighbourhood of G_α on G. Further there exist ∞^2 lines which lie on G and each join P to a point Q of G_β, and each of these lines contains ∞^1 pairs of mates in σ, separating P and Q harmonically. Further, by (v) of § 1, the ∞^2 points Q of G_β joined in this way to P form a Veronese surface ψ_P in G_β with K_P as its ambient space. Hence:

PROPOSITION 2.2. *For $n = 3$, each point P of G_α is joined by a cone of lines of G to a Veronese surface ψ_P of G_β (one of the ∞^3 such surfaces on G_β). The ambient space K_P of ψ_P is the projection from L of the tangent [9] to G at P; and the primal $\mathcal{K}$ generated in M by the ∞^3 spaces K_P represents by projection the whole neighbourhood of G_α on G.*

By Cor. 1.2.1, the Clifford reguli in S_5 – forming a single ∞^{11} system – are represented on G by twisted cubic curves (compounded of the involution σ) which meet each of G_α and G_β in one point; and each of them projects (doubly) from L into a line of M which meets G_β. Since already the lines of M which meet G_β form an irreducible ∞^{11} family, we have

PROPOSITION 2.3. *The ∞^{11} Clifford reguli of S_5 are in birational correspondence with the ∞^{11} lines of M which meet G_β.*

From the above we have a direct confirmation that there exist in S_5 no linear systems of Clifford parallel planes more ample than the Clifford regulus. For any such system would have to be represented in M by a variety (of dimension two or more) such that any general line joining two of its points lies on the variety and is unisecant to G_β; and no such variety exists because G_β, being a quadric Veronesean, possesses no linear subvarieties of positive dimension.

3. The case $n = 4$

Here Ω is a quadric Ω_6^2 of S_7; G is a 16-dimensional variety of order $24,024$ in S_{69}; the two systems of solids on Ω are mapped on G by varieties G_α and G_β each of which (cf. Ch. 5, Prop. 4.1) is the quadratic Veronesean V_6^{128} of a 6-dimensional quadric; the two skew spaces L and M which contain G_α and G_β are each of dimension 34; and the involution σ on G is generated by the chords of G which meet L and M.

By Cor. 1.1.4, G projects doubly from L into a 16-dimensional variety Δ of M, and from M into a 16-dimensional variety Γ of L, each of the two half-Grassmannians Γ, Δ being a birational model of the pairs of polar solids with respect to Ω in S_7.

In the projection of G from L, the exceptional elements arise from (i) the points of G_α, and (ii) the lines which lie on G and join a point of G_α to a point of G_β. Considering these latter first, we observe (by (v) of §1) that the points Q of G_β that are joined to a given point P of G_α by lines lying on G form a threefold ψ_P, quadric Veronesean $V_3^8[9]$ of a solid. More precisely, if we denote by $\sqrt{G_\beta}$ the quadric of which G_β is the quadratic Veronesean, then ψ_P is a member of the ∞^6-family of quadric Veroneseans V_3^8 on G_β that correspond to the generating solids of one system on $\sqrt{G_\beta}$. Further, the ambient [9] of ψ_P is the projection K_P from L of the tangent [16] to G at P. (This [16] meets L in the tangent [6] to G_α at P.) Thus the whole neighbourhood of G_α on G projects from L into a primal (15-dimensional subvariety) of Δ generated by the ∞^6 9-dimensional spaces K_P; and this is the primal we have denoted by $\mathscr{K}$.

We now note further that through any point P of G_α there pass [4]'s which lie entirely on G and that these constitute two triply infinite families, say (λ) and (μ); for if α is the solid of Ω

represented by P, then the solids of S_7 which lie in a fixed [4] through α are mapped on a 4-space λ of G, and the solids of S_7 that meet α in a fixed plane are mapped on a 4-space μ of G. Again (from (vii) of § 1) we note that

(i) the λ and μ generate a cone, vertex P, whose section is a V_6^{20} – Segre product of two solids – and this cone is the model of all the solids of S_7 which meet α in a plane;

(ii) each λ is met by an associated μ in a line of the cone joining P to ψ_P; and

(iii) if PQ is any line of this cone, meeting G_β in Q, then the associated [4]'s λ and μ which meet in this line project from L into the same solid of K_P, namely the tangent solid to ψ_P at Q. Hence:

PROPOSITION 3.1. *For $n = 4$, the Grassmannian G projects doubly from L into the half-Grassmannian Δ, and the total neighbourhood of G_α on G projects into a primal $\mathscr{K}$ of Δ. The variety G_β contains ∞^6 quadric Veronesean threefolds ψ_P, each joined to a point P of G_α by a cone of lines of G. The tangent [16] to G at P projects from L into the ambient space K_P – a [9] – of ψ_P; and the ∞^6 spaces K_P generate the primal $\mathscr{K}$ of Δ. The two systems of [4]'s on G that pass through any point P of G_α project by pairs (one from each system) into the tangent solids of ψ_P.*

The C-curves on G, representing Clifford reguli of S_7, form two ∞^{21}-systems. By Cor. 1.2.1, they are all rational normal quartic curves, those of one system meeting G_α in two points and those of the other system meeting G_β in two points. Those of the first system project doubly from L into lines of Δ – chords of $\mathscr{K}$ – which we call the α-*lines* of Δ; while those of the second system project from L into conics of Δ – each meeting G_β in two points – which we call the β-*conics* of Δ. The same Clifford reguli are represented on the other half-Grassmannian Γ (projection of G from M) by the α-*conics* and β-*lines* of Γ respectively.

Consider now the two types of linear ∞^4-system of Clifford parallel solids in S_7 that we discussed in detail in Ch. 5, §§ 1 and 2, these two types being such that the terminal generators of their Clifford reguli are α-solids in the one case and β-solids in the other.

If $\mathscr{F}$ is a linear ∞^4-system of the first of these types, then it must be represented on Δ by a 4-fold F with the property that if (P, Q) is a generic point-pair of F, then the join of P to Q must be an α-line of Δ and this α-line must lie on F. Hence F is a 4-dimensional linear space lying on Δ. Further, since an α-line meets $\mathscr{K}$ in two points representing the terminal spaces of the corresponding Clifford regulus, it follows that F meets $\mathscr{K}$ in a quadric V_3^2 whose points represent the ∞^3 augmented α-solids that belong to $\mathscr{F}$. Finally, as may easily be verified, the quadric in question is non-singular and does not meet G_β. Recapitulating we have

PROPOSITION 3.2. *For $n = 4$, the two ∞^{21}-systems of Clifford reguli in S_7 are represented on Δ by α-lines, each meeting $\mathscr{K}$ in two points, and by β-conics, each meeting G_β in two points; and similarly on Γ they are represented respectively by α-conics, each meeting G_α in two points, and by β-lines. Further, a linear ∞^4-system of parallel solids of the first kind in S_7 (having α-solids as terminals of its Clifford reguli) is represented on Δ by a space S_4; and this S_4 meets $\mathscr{K}$ in a non-singular quadric threefold representing the augmented α-solids of the system. Similarly, a linear ∞^4-system of the second type in S_7 is represented on Γ by a space Σ_4, and this contains a non-singular quadric threefold representing the augmented β-solids of the system.*

Recalling now that there are ∞^{15} linear ∞^4-systems of each type, we remark that those of the first type, for example, correspond to ∞^{15} [4]'s lying on Δ, that each of these [4]'s contains ∞^6 α-lines, and that the ∞^{21} α-lines so arising account for the whole family of α-lines. This means, in effect, that a Clifford regulus in S_7 belongs to a finite number of linear ∞^4-systems of parallel solids; and in fact it belongs to two such systems (cf. Ch. 5, §2, Ex. 3).

The remaining problem of this section is that of identifying the half-Grassmannians Γ and Δ (which are obviously projectively equivalent varieties) by finding a method of constructing them explicitly. What we find, in fact, is that, exceptionally for the case $n = 4$, each of Γ and Δ is projectively identical with the T-model for $n = 4$; and we prove this by showing that the previous construction of T (cf. Ch. 6, § 3) from its core variety $\mathscr{V}$, which for $n = 4$ is the quadratic Veronesean of a 6-dimensional quadric Ω', is reproduced exactly in a construction of Δ, for example, from its core variety – in this case G_β – which is likewise the quadratic Veronesean of a quadric $\sqrt{G_\beta}$.

Consider then Δ and its core variety G_β. What we have to show is that G_β uniquely determines the aggregate of α-lines of Δ, and therefore also Δ itself since the α-lines generate Δ multiply. As in the previous argument, we note first that if K_P and K_Q are the ambient [9]'s of quadric Veroneseans ψ_P and ψ_Q of G_β (each a $V_3^8[9]$), then there exists a symmetric determinantal Cremona transformation τ between K_P and K_Q, having ψ_P and ψ_Q as its key base varieties, which is such that a pair of points of K_P and K_Q correspond in τ if their join is an α-line. Further, as before, we note that any such transformation with ψ_P and ψ_Q as its key base varieties is uniquely determined by a correlation between ψ_P and ψ_Q – image of a correlation between the two skew generators of $\sqrt{G_\beta}$ of which ψ_P and ψ_Q are the Veronese transforms. We then observe that one such correlation between ψ_P and ψ_Q picks itself out from the others, namely that which corresponds to the natural correlation κ (cf. Ch. 2, § 2) between the two skew generators of $\sqrt{G_\beta}$; and we then see easily that it is precisely this correlation which determines the Cremona transformation τ between K_P and K_Q. Thus, as in the construction of T (Ch. 6, § 3), the core variety G_β of Δ determines the aggregate of α-lines of Δ, and hence Δ itself. This proves

PROPOSITION 3.3. *For $n = 4$, each of the half-Grassmannians Γ and Δ is projectively equivalent to the corresponding T-model of the pairs of polar solids with respect to Ω in S_7.*

Although Γ, Δ and T are projectively equivalent to one another, the pairs of polar solids in S_7 are mapped on their points in very different ways; and in the following section we shall be considering the resulting (non-linear) birational correspondences between pairs of the three models in question. In preparation for this we need the following further property of the representation on Δ:

PROPOSITION 3.4. *There exists on* Δ, *in addition to its* ∞^{21}-*system of* α-*lines, a second* ∞^{21}-*system of lines – to be called the* γ-*lines of* Δ – *each of which represents a pair of pencils of solids in* S_7 *such that the solids of one pencil are polars of those of the other for* Ω. *Further these* γ-*lines (like the* α-*lines) lie in* ∞^{15} [4]'s *of a second system on* Δ.

Proof. It is clear, first of all, that any pencil of solids of S_7 is represented on G by a line l and that the polar pencil of solids will be represented on G by a second line $\sigma(l)$, image of l under the involution σ. Moreover, in the projection of G from L onto Δ, these two lines must project into the same line – a γ-line – of Δ.

Further, if π is any general plane of S_7, then the solids of S_7 through π are mapped on G by the points of a space S_4; and similarly, the solids of S_7 that lie in the polar [4] of π are mapped on G by the points of another space S_4'. Plainly then, in the projection from L, the two spaces S_4 and S_4' project into the same [4] of Δ; and there are altogether ∞^{15} such [4]'s on Δ, corresponding to the ∞^{15} planes π of S_7. Each such [4] contains ∞^6 γ-lines representing polar pairs of pencils of S_7; and this proves the result.

COROLLARY 3.4.1. *There exists, similarly, a second system of lines on* Γ *(besides the* β-*lines); and, since these represent the same polar pairs of pencils of* S_7 *as the* γ-*lines of* Δ, *we shall call them the* γ-*lines of* Γ.

4. Construction and extension of the Study triality in S_7

For the following discussion it will be convenient, as regards the two types of Clifford regulus in S_7, to revert to a notation already introduced in Ch. 3, §2. A Clifford regulus whose terminal solids are both α-solids of Ω will therefore be called an α-*regulus;* and one whose terminals are β-solids of Ω will be called a β-*regulus*.

In the preceding section we obtained two birational mappings of the polar pairs (Π, Π') of solids of S_7 with respect to Ω, namely, that in which the (Π, Π') are mapped on the points of a half-Grassmannian Δ by the double projection of G from L into M, and that in which they are mapped on the points of Γ by the double projection of G from M into L.

As regards Δ, we saw that this model was completely determined by its core variety G_β, quadratic Veronesean of a 6-dimensional quadric which we denoted by $\sqrt{G_\beta}$. In particular:

(i) There exist on G_β two ∞^6-systems of quadric Veroneseans $V_3^8[9]$, to be called V_β and W_β respectively, which correspond to the two systems of generating solids of $\sqrt{G_\beta}$, to be denoted by $\sqrt{V_\beta}$ and $\sqrt{W_\beta}$ respectively. We take the V_β to be those Veroneseans which were previously called the ψ_P, recalling (from §3) that each of them is joined to a point P of G_α by a cone of lines lying on G, and that it represents in S_7 the ∞^3 β-solids of Ω which meet a given α-solid in a plane. The W_β then each represent in S_7 the ∞^3 β-solids that pass through a fixed point of Ω.

(ii) There exist on Δ two primals (15-dimensional subvarieties), $\mathscr{K}$ and $\mathscr{J}$, such that $\mathscr{K}$ is generated by the ∞^6 ambient (9-dimensional) spaces K_β of the V_β, while $\mathscr{J}$ is generated by the ∞^6 ambient spaces J_β of the W_β. The points of $\mathscr{K}$ represent, in general, the augmented generators $\alpha(q)$ of Ω in S_7, those of them which lie in a space K_β representing the $\alpha(q)$ for which α is fixed (cf. §3, where the notation K_P was used for a K_β).

(iii) There exist on Δ two ∞^{21}-systems of lines (each generating Δ multiply); and these are, respectively, the α-lines of Δ, each representing an α-regulus in S_7, and the γ-lines, each of which represents the polar pairs (Π, Π') of two mutually polar pencils of solids in S_7. The α-lines are the joins of corresponding points in Cremona transformations τ (determined by G_β) between every pair of the spaces K_β; and similarly the γ-lines are the joins of corresponding points in other Cremona transformations τ' (also determined by G_β) between pairs of the spaces J_β.

As regards Γ and its core variety G_α, similar properties hold. In particular, G_α has systems of Veroneseans V_α and W_α which correspond to the generating solids $\sqrt{V_\alpha}$ and $\sqrt{W_\alpha}$ of $\sqrt{G_\alpha}$; and each V_α represents the α-solids of Ω which meet a fixed β-solid in a plane, while each W_α represents the α-solids which pass through a fixed point of Ω. Also Γ possesses two ∞^{21}-systems of lines, namely its β-lines which represent β-reguli in S_7, and its γ-lines which each again represent the polar pairs (Π, Π') of two mutually polar pencils of solids of S_7.

We now concentrate our attention, in the first instance, on the relations between the three varieties Ω, G_α and G_β. As between Ω and G_β, we note the following correspondences:

β-solid of Ω $\longleftrightarrow$ point of G_β,

α-solid of Ω (as defining the $\longleftrightarrow$ V_β of G_β, aggregate of β-solids which meet it in a plane)

point of Ω (as defining the $\longleftrightarrow$ W_β of G_β. aggregate of β-solids which pass through it)

More succinctly, if we replace G_β by the quadric $\sqrt{G_\beta}$ from which it is derived, the above correspondences take the form

$$\left.\begin{array}{ccc} \beta\text{-solid of } \Omega & \longleftrightarrow & \text{point of } \sqrt{G_\beta}, \\ \alpha\text{-solid of } \Omega & \longleftrightarrow & \text{solid } \sqrt{V_\beta} \text{ of } \sqrt{G_\beta}, \\ \text{point of } \Omega & \longleftrightarrow & \text{solid } \sqrt{W_\beta} \text{ of } \sqrt{G_\beta}. \end{array}\right\} \quad (4.1)$$

Similarly, as between Ω and $\sqrt{G_\alpha}$, we have the correspondences

$$\left.\begin{array}{ccc} \alpha\text{-solid of } \Omega & \longleftrightarrow & \text{point of } \sqrt{G_\alpha}, \\ \beta\text{-solid of } \Omega & \longleftrightarrow & \text{solid } \sqrt{V_\alpha} \text{ of } \sqrt{G_\alpha}, \\ \text{point of } \Omega & \longleftrightarrow & \text{solid } \sqrt{W_\alpha} \text{ of } \sqrt{G_\alpha}. \end{array}\right\} \quad (4.2)$$

Finally, as between G_α and G_β, we make two points correspond if their join lies on G; and this induces a correspondence between $\sqrt{G_\alpha}$ and $\sqrt{G_\beta}$ in which

$$\left.\begin{array}{ccc} \text{point of } \sqrt{G_\alpha} & \longleftrightarrow & \text{solid } \sqrt{V_\beta} \text{ of } \sqrt{G_\beta}, \\ \text{solid } \sqrt{V_\alpha} \text{ of } \sqrt{G_\alpha} & \longleftrightarrow & \text{point of } \sqrt{G_\beta}, \\ \text{solid } \sqrt{W_\alpha} \text{ of } \sqrt{G_\alpha} & \longleftrightarrow & \text{solid } \sqrt{W_\beta} \text{ of } \sqrt{G_\beta}, \end{array}\right\} \quad (4.3)$$

where, in this last entry, we mean that a solid $\sqrt{W_\alpha}$ of $\sqrt{G_\alpha}$ corresponds to a solid $\sqrt{W_\beta}$ of $\sqrt{G_\beta}$ if they represent the same point of Ω in the correspondences (4.1) and (4.2).

We have thus derived geometrically the system consisting of three 6-dimensional quadrics, in mutual correspondence, which is known as the Study triality.†

† There is an extensive literature dealing with the Study triality, and the reader interested in this topic may consult the bibliographies of two recent papers on the subject: Tits[25] and Dye[8].

The principal property of the triality is that each of the three correspondences preserves *incidence*, where incidence is defined as follows:

(*a*) a point is incident with a solid if it lies in the solid;

(*b*) two points of a (6-dimensional) quadric are incident if they are conjugate points, i.e. if their join lies on the quadric;

(*c*) two solids of the same system on a quadric are incident if they meet in a line; and

(*d*) two solids of opposite systems on a quadric are incident if they meet in a plane.

The proof that all these incidence relations are preserved under the triality – and other known results besides – follow very easily in terms of our model, and we leave the reader to fill in the details.

We now recall the interesting coincidence (Prop. 3.3) that in the present case ($n = 4$) the T-model is projectively the same variety as each of the half-Grassmannians Γ and Δ. If $\mathscr{V}$ denotes the core variety of T, then $\mathscr{V}$ is the quadratic Veronesean of a 6-dimensional quadric that we originally denoted by Ω' (Ch. 6, §3) but which we may now denote for consistency by $\sqrt{\mathscr{V}}$. We also denote the two systems of Veroneseans $V_3^8[9]$ on $\mathscr{V}$ by ϕ and ψ, remembering that these correspond to generating solids of $\sqrt{\mathscr{V}}$ which we denote by $\sqrt{\phi}$ and $\sqrt{\psi}$. We recall, then, that the ambient spaces A of the Veroneseans ϕ generate a primal $\mathscr{A}$ of T, whose points represent in general the augmented generators $\alpha(q)$ of Ω in S_7, and, similarly, the ambient spaces B of the ψ generate a primal $\mathscr{B}$ of T whose points represent in general the augmented generators $\beta(q)$. Also T contains two ∞^{21}-systems of lines, its α-lines representing α-reguli and its β-lines representing β-reguli, such that each of the former meets a pair of the spaces A in points which correspond in a Cremona transformation (determined by $\mathscr{V}$) between the spaces; while the β-lines are similarly defined.

We now make points of Γ, Δ and T correspond to each other if they represent the same polar pair in S_7, and we thereby obtain three commuting birational correspondences λ, μ, ν as indicated in the diagram:

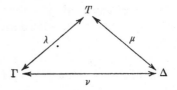

These correspondences, of course, are not trivial (i.e. not collineations); for an α-regulus is represented on each of T and Δ by a line, but on Γ by a conic; whereas a β-regulus is represented on each of T and Γ by a line, but on Δ by a conic. Further, it is easy to verify that a system of polar pairs of the kind that is represented on Γ or Δ by a γ-line is represented on T by a conic. Hence:

PROPOSITION 4.1. *There exist on each of the (projectively equivalent) varieties Γ, Δ and T three ∞^{21}-systems of curves (either lines or conics), representing α-reguli, β-reguli and pairs of mutually polar pencils of solids, respectively, in S_7. The birational correspondences (induced by the representations) between pairs of Γ, Δ and T transform the triads of curve systems in question according to the columns of the table.*

	α-regulus	β-regulus	Pair of mutually polar pencils of solids
Γ	α-conic	β-line	γ-line
Δ	α-line	β-conic	γ-line
T	α-line	β-line	γ-conic

As regards the three birational correspondences λ, μ, ν, it is clear that each of the three core varieties, G_α, G_β and $\mathscr{V}$, is fundamental for those two of λ, μ, ν that carry the corresponding variety Γ, Δ or T into either of the other two, transforming it in each case into a primal of the other. We note, in fact

COROLLARY 4.1.1. *The restriction of each of the birational correspondences λ, μ, ν to the core varieties is an ∞^9-correspondence, and these ∞^9-correspondences induce, between the quadrics $\sqrt{G_\alpha}$, $\sqrt{G_\beta}$ and $\sqrt{\mathscr{V}}$, the three correspondences of a Study triality.*

THE HURWITZ–RADON MATRIX EQUATIONS

As we have shown in Ch. 4, the existence of an r-dimensional non-singular linear system of Clifford parallels in S_{2n-1} $(n \geqslant 2)$ is equivalent, algebraically, to the existence of a set of $r-1$ (complex) matrices $S_1, \ldots, S_{r-1}$ of order n satisfying the Hurwitz–Radon matrix equations

$$\left. \begin{aligned} S_i^T &= -S_i, \quad S_i^2 = -I \quad (i = 1, \ldots, r-1), \\ S_i S_j &= -S_j S_i \quad (i, j = 1, \ldots, r-1; \; i \neq j). \end{aligned} \right\} \tag{0.1}$$

The object of this appendix is to establish the following

THEOREM A 1. *If n is written in the form $n = u \cdot 2^{4\alpha + \beta}$ (with u odd and $0 \leqslant \beta \leqslant 3$) and if $H(n) = 8\alpha + 2^\beta$, then the equations (0.1) admit solutions in matrices of order n if and only if $r \leqslant H(n)$. In particular, solutions of (0.1), with $r = n$, exist only in the three cases $n = 2, 4$ and 8.*

It will appear, incidentally, that there are, up to orthogonal equivalence, only a finite number of solution sets of (0.1) for given n and r; so that, by Ch. 4, Cor. 3.1.2, there are only a finite number of projectively distinct non-singular linear systems of Clifford parallels in S_{2n-1}.

The proof of Theorem A 1 will be given in § 2 below. Before this, however, we include some brief historical notes by way of introduction.

1. Historical notes

The equations (0.1) appeared first in the literature in connection with the problem of composition of quadratic forms, which may be stated as follows.

Find the values of n and r for which there exist n forms $z_1, ..., z_n$, bilinear with complex coefficients in the two sets of indeterminates $x_1, ..., x_r$ and $y_1, ..., y_n$, satisfying the identity

$$(x_1^2 + ... + x_r^2)(y_1^2 + ... + y_n^2) \equiv z_1^2 + ... + z_n^2. \qquad (1.1)$$

This problem, which generalizes the famous eight squares problem, was first proposed and solved by Hurwitz in his posthumous paper [15]. The same author had solved the problem in the case $n = r$ in an earlier paper [14].

To see how this problem reduces to the equations (0.1), suppose that

$$z_i = \sum_{j=1}^{n} \sum_{h=1}^{r} a_{ihj} x_h y_j \quad (i = 1, ..., n), \qquad (1.2)$$

and, for $h = 1, ..., r$, let A_h denote the matrix of order n whose (i,j) element is a_{ihj}. If, then, we write

$$\Lambda = x_1 A_1 + ... + x_r A_r,$$

we see that the (i,j) element of Λ is

$$\lambda_{ij} = \sum_{h=1}^{r} x_h a_{ihj}$$

and (1.2) may be written

$$z_i = \sum_{j=1}^{n} \lambda_{ij} y_j \quad (i = 1, ..., n).$$

Hence, using x, y, z to denote the column vectors (x_i), (y_i), (z_i), we have $z = \Lambda y$ and the desired identity (1.1) becomes

$$(x^T x)(y^T y) \equiv z^T z \equiv y^T \Lambda^T \Lambda y;$$

this is plainly possible if and only if $\Lambda^T \Lambda \equiv (x^T x) I$ or, in full, if and only if

$$(x_1 A_1^T + ... + x_r A_r^T)(x_1 A_1 + ... + x_r A_r) \equiv (x_1^2 + ... + x_r^2) I.$$

Thus, comparing coefficients on either side of this identity, we require

$$A_i^T A_i = I \quad (i = 1, ..., r)$$

and

$$A_i^T A_j = -A_j^T A_i \quad (i, j = 1, ..., r; i \neq j).$$

Finally, writing $A_i = A_r S_i$ ($i = 1, ..., r-1$), these last equations reduce precisely to the form (0.1), together with $A_r^T A_r = I$.

Contemporaneously with Hurwitz, Radon [20] showed that the inequality of Theorem A 1 is necessary and sufficient for the existence of *real* solutions of (0.1). This means, of course, that whenever equations (0.1) admit complex solutions, then they necessarily also admit real solutions; this fact was made particularly clear in a subsequent proof of Theorem A 1 by Eckmann[9]. Another proof of Radon's result was given by Lee [18]. In another direction, Albert[2] and Dubisch[7] solved the composition problem for quadratic forms over more general fields, and later authors have generalized and extended their work. Finally, with a slightly different emphasis, Wong[27] discussed the equations (0.1) with particular reference to the *maximal real* solutions (i.e. the solutions of (0.1) that cannot be extended† to solutions involving more than $r - 1$ matrices).

No account of the equations (0.1) would be complete without mention of their close connection with a well-known problem concerning vector fields. We recall that a *q-field* on a smooth manifold M is a set of q smooth tangent vector fields on M such that, at every point of M, the vectors belonging to the q fields are linearly independent. An obviously important question is that of determining the maximum possible q for a given manifold M.

Let us say that a set of p real matrices $A_1, ..., A_p$ of order n is a *strongly linearly independent set* if, for all non-zero real vectors x, the vectors
$$A_1 x, ..., A_p x$$
are linearly independent. (This is equivalent to saying that 0 is the only singular matrix in the real linear family
$$\lambda_1 A_1 + ... + \lambda_p A_p.)$$
Now Stiefel[24] proved that, if there exists a strongly linearly independent set of p matrices of order n, then there exists a $(p-1)$-field on the real projective $(n-1)$-space, and consequently also on the $(n-1)$-sphere. However, if $S_1, ..., S_{r-1}$ is a set of real matrices satisfying (0.1) and if $B = \alpha I + \lambda_1 S_1 + ... + \lambda_{r-1} S_{r-1}$, then
$$B^T B = (\alpha^2 + \lambda_1^2 + ... + \lambda_{r-1}^2) I,$$
from which it follows that $I, S_1, ..., S_{r-1}$ is a strongly linearly independent set. We deduce that, *if the integers r and n are such that the equations* (0.1) *admit solutions, then there exists an $(r-1)$-field on real projective $(n-1)$-space and on the $(n-1)$-sphere.*

† Cf. Ch. 4, §5, Ex. 2.

More recently, Adams[1] has solved the converse problem of determining all cases in which an $(r-1)$-field may exist on an $(n-1)$-sphere. Using powerful techniques of K-theory, he proves that *such fields can exist if and only if the inequality of Theorem* A 1 *is satisfied.* A brief account of Adams' work, and of a number of other related questions, is given by Atiyah[3]. The reader may also consult Husemoller[16], particularly the chapter on Clifford algebras.

2. Eckmann's proof of Theorem A1

The proof to be given here of Theorem A 1 follows closely the account given by Eckmann[9], which was inspired by a proof for the special case $n = r$ given by Jordan, von Neumann and Wigner[17]. In addition to being quite short, it provides an elegant and simple example of the application of general theorems from the representation theory of finite groups (in contrast to the proofs of Hurwitz and Radon, which employ largely *ad hoc* methods, outside the mainstream of modern algebra). The results needed from group representation theory are elementary and may be found in Serre[22].

We observe, first of all, that a skew-symmetric matrix S_i satisfies $S_i^2 = -I$ if and only if it is (complex) orthogonal. We are concerned, then, with the problem of determining values of r and n for which there exist orthogonal matrices $S_1, ..., S_{r-1}$ of order n satisfying

$$\left.\begin{array}{l} S_i^2 = -I \quad (i = 1, ..., r-1), \\ S_i S_j = -S_j S_i \ (i,j = 1, ..., r-1; \ i \neq j). \end{array}\right\} \qquad (2.1)$$

Let G denote the abstract group on r generators $e, s_1, ..., s_{r-1}$ subject to the relations†

$$\left.\begin{array}{l} e^2 = 1, \\ s_i^2 = e, \quad s_i e = e s_i \quad (i = 1, ..., r-1), \\ s_i s_j = e s_j s_i \quad (i,j = 1, ..., r-1; \ i \neq j). \end{array}\right\} \qquad (2.2)$$

† We present the relations for G in the form given by Eckmann[9]; but it may be noted that the relations $s_i e = e s_i$ can be omitted (being consequences of $s_i^2 = e$).

Our problem, as just stated, is plainly equivalent to that of determining whether there exists a representation of G by (complex) orthogonal matrices of order n, with e represented by the matrix $-I$. To solve the problem in this form we shall regard r as fixed and proceed in four steps:

(A) we first find the degrees of the *irreducible* representations of G in which e is represented by $-I$ (ignoring temporarily the requirement of orthogonality);

(B) we next find the degrees of *all* representations of G in which e is represented by $-I$;

(C) we then decide which of the representations found in (B) are equivalent to orthogonal representations; and finally

(D) we find that an orthogonal representation, in which e is represented by $-I$, can be of degree n if and only if the inequality of Theorem A 1 is satisfied.

(A) To begin with, it follows from the relations (2.2) that all the elements of G are of the form

$$s_{i_1} s_{i_2} \ldots s_{i_k} \quad \text{or} \quad e s_{i_1} s_{i_2} \ldots s_{i_k}, \tag{2.3}$$

where $0 \leqslant k \leqslant r-1$ and the integers $i_1, i_2, \ldots, i_k$ take all possible sequences of values from $1, \ldots, r-1$ such that $i_1 < i_2 < \ldots < i_k$. The order of G is therefore 2^r, and an explicit list of its elements is

$$\left. \begin{array}{l} 1; s_1, \ldots, s_{r-1}; s_1 s_2, \ldots, s_{r-2} s_{r-1}; \ldots; s_1 s_2 \ldots s_{r-1}; \\ e; e s_1, \ldots, e s_{r-1}; e s_1 s_2, \ldots, e s_{r-2} s_{r-1}; \ldots; e s_1 s_2 \ldots s_{r-1}. \end{array} \right\} \tag{2.4}$$

When $r = 2$, G is the cyclic group of order four and this, as is well-known, possesses just four irreducible representations (each of degree one). *We assume from now on that $r \geqslant 3$.*

It is easy to verify that each commutator of G is either 1 or e; so the commutator subgroup K of G is of order two and the abelian group $G' = G/K$ is of order 2^{r-1}. Hence G *possesses 2^{r-1} inequivalent representations of degree one.*

It is again easy to verify that a complete set of conjugates of an element g of G consists either of g alone or of g and eg. More precisely:

(a) If r is odd, then 1 and e are conjugate only to themselves, while every other element g is conjugate to eg.

(b) If r is even, then $1, e, s_1 s_2 \ldots s_{r-1}$ and $e s_1 s_2 \ldots s_{r-1}$ are conjugate to themselves, while every other element g is conjugate to eg.

Thus, letting h denote the number of conjugacy classes in G, we have:

(a) If r is odd, there are two conjugacy classes with one element and $\frac{1}{2}(2^r - 2)$ classes with two elements, so that $h = 2^{r-1} + 1$.

(b) If r is even, there are four conjugacy classes with one element and $\frac{1}{2}(2^r - 4)$ classes with two elements, so that $h = 2^{r-1} + 2$.

Since the number of inequivalent irreducible representations of G is h, we have (besides the 2^{r-1} representations of degree one noted earlier)

(a) if r is odd, just one irreducible representation of degree $d > 1$,

(b) if r is even, just two irreducible representations of degrees d, d' ($d > 1, d' > 1$).

Moreover, since the order of a group is equal to the sum of the squares of the degrees of its irreducible representations, the values of d and d' are easily calculated; namely

(a) if r is odd, then $d^2 + 2^{r-1} = 2^r$, so that $d = 2^{\frac{1}{2}(r-1)}$;

(b) if r is even, then $d^2 + d'^2 + 2^{r-1} = 2^r$, so that

$$d^2 + d'^2 = 2^{r-1};$$

and, remembering that d and d' must divide the order 2^r of G, we must have $d = 2^c, d' = 2^{c'}$ and hence

$$2^{2c} + 2^{2c'} = 2^{r-1},$$

from which it follows that $c = c' = \frac{1}{2}(r - 2)$ and $d = d' = 2^{\frac{1}{2}(r-2)}$.

In the 2^{r-1} representations of G of degree one, e is represented by the unit matrix $I = (1)$. On the other hand, since the group $G' = G/(1, e)$ is abelian, e cannot be represented by the unit matrix in the irreducible representations of higher degree. But, since e lies in the centre of G, its corresponding matrix (which must commute with all the matrices of the representation) is necessarily a multiple λI of the unit matrix; and, since $e^2 = 1$, we must have $\lambda^2 = 1$, so that $\lambda = -1$ and e is represented by $-I$. Hence:

There is, for r odd, just one irreducible representation of G (up to equivalence) which satisfies our condition that e be represented by $-I$, and it is of degree $2^{\frac{1}{2}(r-1)}$; while, for n even, there are two such representations, each of degree $2^{\frac{1}{2}(r-2)}$. The latter statement is valid also when $r = 2$, a case we excluded earlier.

(B) It follows now that the degree n of any (not necessarily irreducible) representation of G, in which e is represented by $-I$, is of the form

(a) $n = m \cdot 2^{\frac{1}{2}(r-1)}$ if r is odd,

(b) $n = m \cdot 2^{\frac{1}{2}(r-2)}$ if r is even,

where m is a positive integer. In case (a) there is, up to equivalence, a unique representation (of the required kind) of degree n; whereas, in case (b), there are $m + 1$ inequivalent representations of degree n. The two cases may be conveniently summed up in the following statement:

If $n = u \cdot 2^t$, where u is odd, then there exists a representation of G of degree n, with e represented by $-I$, if and only if $r \leqslant 2t + 2$.

(C) We now proceed to pick out, from among the representations of G that we have found, those which are equivalent to orthogonal representations. These, by a result of Frobenius and Schur, are precisely those which are equivalent to real representations (and hence also to real orthogonal representations). First, then, we must discover, from among the *irreducible* representations that we have found, those which are equivalent to real representations.

To this end, we consider the sum

$$F = \sum_{g \in G} \chi(g^2)$$

of the characters† of the squares of all the elements $g \in G$, recalling that an irreducible representation D is (i) equivalent to a real one if and only if the character-sum F is positive, (ii) not equivalent to a real representation, but equivalent to its complex conjugate representation $\bar{D}$, if F is negative, and (iii) not equivalent to a real representation, nor to $\bar{D}$, if $F = 0$.

Now (cf. equation (2.3)), any element $g \in G$ is of the form

$$g = s_{i_1} s_{i_2} \ldots s_{i_k} \quad \text{or} \quad g = e s_{i_1} s_{i_2} \ldots s_{i_k},$$

---

† The character of an element $g \in G$, in a given representation of G, is the trace of the matrix corresponding to g.

and in either case we find easily, by use of (2.2), that

$$g^2 = e^{\frac{1}{2}k(k+1)},$$

so that $\qquad g^2 = 1$ if $k \equiv 3$ or $0 \pmod 4$

and $\qquad g^2 = e$ if $k \equiv 1$ or $2 \pmod 4$.

Hence, if a representation D of G carries e into $-I$ and is of degree d, we have

$$\chi(g^2) = d \qquad \text{if} \quad k \equiv 3 \text{ or } 0 \pmod 4$$

and $\qquad \chi(g^2) = -d$ if $k \equiv 1$ or $2 \pmod 4$.

Thus, summing over the 2^r elements of G given by (2.4), we have

$$F = \sum_{g \in G} \chi(g^2) = 2d\sigma,$$

where σ denotes the sum

$$\sigma = \binom{r-1}{0} - \binom{r-1}{1} - \binom{r-1}{2} + \binom{r-1}{3} + \binom{r-1}{4} - \cdots \pm \binom{r-1}{r-1}.$$

To determine the sign of σ, we consider the complex number

$$z = (1-i)^{r-1} = x + iy \quad (x, y \text{ real}),$$

noting that σ is just $x + y$. Thus, since $\arg z = -\frac{1}{4}\pi(r-1)$, we have the following cases:

$$\sigma > 0 \quad \text{if} \quad -\frac{\pi}{4}(r-1) \equiv 0, \ \frac{\pi}{4} \ \text{or} \ \frac{2\pi}{4} \pmod{2\pi},$$

$$\sigma = 0 \quad \text{if} \quad -\frac{\pi}{4}(r-1) \equiv \frac{3\pi}{4} \ \text{or} \ \frac{7\pi}{4} \pmod{2\pi},$$

$$\sigma < 0 \quad \text{if} \quad -\frac{\pi}{4}(r-1) \equiv \frac{4\pi}{4}, \ \frac{5\pi}{4} \ \text{or} \ \frac{6\pi}{4} \pmod{2\pi}.$$

It follows that F is positive if $r \equiv 7$, 0 or 1 $\pmod 8$, negative if $r \equiv 3$, 4 or 5 $\pmod 8$, and zero if $r \equiv 2$ or 6 $\pmod 8$. Hence:

The irreducible representations of G found earlier are equivalent to real representations when $r \equiv 7$, 0 or 1 (mod 8), but not for other values of r. When $r \equiv 3$, 4 or 5 (mod 8), they are equivalent to their complex conjugate representations, but not so when $r \equiv 2$ or 6

$(mod\,8)$. *When* $r \equiv 2$ *or* 6 $(mod\,8)$, *each of the two irreducible representations is equivalent to the complex conjugate of the other.*

It follows now that the degree n of a real representation of G, in which e is represented by $-I$, must be an even multiple of the degree of the irreducible representation(s),† except when $r \equiv 7$, 0 or 1 (mod 8). The degree of a real-irreducible representation must therefore be equal to

$$
\left.
\begin{array}{ll}
(a) \; 2^{\frac{1}{2}(r-1)} & \text{if} \quad r \equiv 7 \text{ or } 1 \quad (\text{mod } 8), \\
(b) \; 2^{\frac{1}{2}(r-2)} & \text{if} \quad r \equiv 0 \quad (\text{mod } 8), \\
(c) \; 2 \cdot 2^{\frac{1}{2}(r-1)} = 2^{\frac{1}{2}(r+1)} & \text{if} \quad r \equiv 3 \text{ or } 5 \quad (\text{mod } 8), \\
(d) \; 2 \cdot 2^{\frac{1}{2}(r-2)} = 2^{\frac{1}{2}r} & \text{if} \quad r \equiv 2, 4 \text{ or } 6 \quad (\text{mod } 8).
\end{array}
\right\} \quad (2.5)
$$

Moreover there are, up to equivalence, two such representations when $r \equiv 0$ or 4 (mod 8), while there is only one for other values of r. Hence:

Orthogonal representations of G *of degree* $n = m \cdot 2^q$ *(where* m *is a positive integer), in which* e *is represented by* $-I$, *exist only in the following cases:*

$$
\left.
\begin{array}{lllll}
(a) & q \equiv 3 \text{ or } 0 & (\text{mod } 4) & and & r = 2q+1, \\
(b) & q \equiv 3 & (\text{mod } 4) & and & r = 2q+2, \\
(c) & q \equiv 2 \text{ or } 3 & (\text{mod } 4) & and & r = 2q-1, \\
(d) & q \equiv 1, 2 \text{ or } 3 & (\text{mod } 4) & and & r = 2q.
\end{array}
\right\} \quad (2.6)
$$

(D) So far, we have been thinking of r as fixed and have found all possible values of n for which the required representations exist. If we now reverse this emphasis, thinking of n as given and looking for the possible values of r, we see easily that (2.6) is equivalent to the following:

If $n = u \cdot 2^t$ *(with* u *odd), then the largest possible value* $H(n)$ *of* r *is given by*

$$
\begin{array}{ll}
H(n) = 8\alpha + 1 & when \quad t = 4\alpha, \\
H(n) = 8\alpha + 2 & when \quad t = 4\alpha + 1, \\
H(n) = 8\alpha + 4 & when \quad t = 4\alpha + 2, \\
H(n) = 8\alpha + 8 & when \quad t = 4\alpha + 3.
\end{array}
$$

† A real representation of least degree, for $r = 2, \ldots, 6$ (mod 8), is equivalent to the direct sum of an irreducible representation with its conjugate complex representation.

Combining these four possibilities into one, namely,

$$H(n) = 8\alpha + 2^\beta \quad \text{when} \quad t = 4\alpha + \beta \quad (0 \leqslant \beta \leqslant 3),$$

we have finally the result of Theorem A 1.

If r satisfies the condition $r \leqslant H(n)$ then, by the note following equations (2.5) above, there is exactly one solution (up to equivalence) of the equations (0.1) when $r \not\equiv 0 \pmod 4$; whereas, if $r \equiv 0 \pmod 4$, there are $m + 1$ different solutions, where

$$m = \frac{n}{2^{\frac{1}{2}(r-2)}} \quad \text{when} \quad r \equiv 0 \pmod 8,$$

and

$$m = \frac{n}{2^{\frac{1}{2}r}} \quad \text{when} \quad r \equiv 4 \pmod 8.$$

The explicit solutions, in all cases, may easily be deduced from the forms given by Hurwitz [15], although this author takes the original equations (0.1) in a slightly different form.

Finally, we again draw attention to the fact (already noted in Ch. 4, §5, Ex. 2) that a solution of (0.1), with $r < H(n)$, is not necessarily extendable to a solution with $r = H(n)$.

3. Explicit solutions for the cases n = 2, 4 and 8

For completeness, we give here the solutions (up to orthogonal equivalence) of the equations (0.1) in the three cases of most interest, $n = 2$, 4 and 8.

We first define matrices I, J, K, L of order two as follows:

$$I = \begin{pmatrix} 1 & 0 \\ 0 & 1 \end{pmatrix}, \quad J = \begin{pmatrix} 0 & -1 \\ 1 & 0 \end{pmatrix}, \quad K = \begin{pmatrix} 1 & 0 \\ 0 & -1 \end{pmatrix}, \quad L = \begin{pmatrix} 0 & 1 \\ 1 & 0 \end{pmatrix},$$

noting that

$$J^2 = -I, \quad K^2 = L^2 = I,$$

$$KL = -LK = -J,$$

$$LJ = -JL = K,$$

$$JK = -KJ = L.$$

(i) If $n = 2$, then $H(n) = 8\alpha + 2^\beta = 2$ and the only solutions of $S_1^2 = -I$ are $S_1 = J$ and $S_1 = -J$. The two solutions are orthogonally equivalent.

(ii) If $n = 4$, then $H(n) = 4$. One solution of (0.1) with $r = 4$ is†

$$S_1' = I \times J, \quad S_2' = J \times K, \quad S_3' = J \times L$$

and every solution with $r = 4$ is orthogonally equivalent either to (S_1', S_2', S_3') or to $(-S_1', -S_2', -S_3')$. The solution of (0.1) with $r = 3$ (resp. $r = 2$) is unique up to equivalence, an explicit solution being (S_1', S_2') (resp. (S_1')).

(iii) If $n = 8$, then $H(n) = 8$. One solution of (0.1) with $r = 8$ is

$$S_1'' = I \times I \times J, \quad S_2'' = I \times J \times K,$$
$$S_3'' = J \times I \times L, \quad S_4'' = J \times K \times K,$$
$$S_5'' = J \times L \times K, \quad S_6'' = K \times J \times L,$$
$$S_7'' = L \times J \times L,$$

and every solution with $r = 8$ is orthogonally equivalent either to $(S_1'', ..., S_7'')$ or to $(-S_1'', ..., -S_7'')$. If $r < 8$ and $r \neq 4$, the solution of (0.1) is unique up to equivalence, an explicit solution being $(S_1'', ..., S_{r-1}'')$. If $r = 4$, however, there are three inequivalent solutions, one of which may be taken to be (S_1'', S_2'', S_3''); the other two are, modulo equivalence,

$$I \times I \times J, \quad I \times J \times K, \quad I \times J \times L$$

and its negative. These two 'extra' solutions are not extendable to solutions involving more than three matrices; the geometrical significance of this has already be explained (Ch. 5, §4).

† The cross denotes Kronecker product of matrices, and the solutions may be easily checked by use of the basic multiplication rule

$$(A \times B)(C \times D) = AC \times BD.$$

REFERENCES

[1] Adams, J. F. Vector fields on spheres. *Annals of Math.* (2) **75** (1962), 603–632.

[2] Albert, A. A. Quadratic forms permitting composition. *Annals of Math.* (2) **43** (1942), 161–177.

[3] Atiyah, M. F. The role of algebraic topology in mathematics. *J. London Math. Soc.* **41** (1966), 63–69.

[4] Burau, W. *Mehrdimensionale projektive und höhere Geometrie* (Berlin, 1961).

[5] Clifford, W. K. Preliminary sketch of biquaternions. *Proc. London Math. Soc.* (1) **4** (1873), 381–395. Reprinted in *Mathematical Papers*, 181–200.

[6] Coxeter, H. S. M. *Non-Euclidean geometry* (Toronto, 1942).

[7] Dubisch, R. Composition of quadratic forms. *Annals of Math.* (2) **47** (1946), 510–527.

[8] Dye, R. H. The simple group FH(8, 2) of order $2^{12}.3^5.5^2.7$ and the associated geometry of triality. *Proc. London Math. Soc.* (3) **18** (1968), 521–562.

[9] Eckmann, B. Gruppentheoretischer Beweis des Satzes von Hurwitz–Radon über die Komposition quadratischer Formen. *Comment. Math. Helv.* **15** (1943), 358–366.

[10] Forsyth, A. R. *Geometry of four dimensions*, Vol. 1 (Cambridge University Press, 1930), particularly Chapter VI.

[11] Heymans, P. Pfaffians and skew-symmetric matrices. *Proc. London Math. Soc.* (3) **19** (1969), 730–768.

[12] Heymans, P. The minimum model of spaces on a quadric (to appear in *Proc. London Math. Soc.*).

[13] Hodge, W. V. D. and Pedoe, D. *Methods of algebraic geometry*, Vols. 1 and 2 (Cambridge University Press, 1947 and 1952).

[14] Hurwitz, A. Über die Komposition der quadratischen Formen von beliebig vielen Variablen. *Nach. v. der Ges. der Wiss. Göttingen, Math.-Phys. Klasse* (1898), 309–316. Reprinted in *Math. Werke, Bd. 2*, 565–571.

[15] Hurwitz, A. Über die Komposition der quadratischen Formen. *Math. Ann.* **88** (1923), 1–25. Reprinted in *Math. Werke, Bd. 2*, 641–666.

[16] Husemoller, D. *Fibre bundles* (McGraw-Hill, New York, 1966).

[17] Jordan, P., Von Neumann, J. and Wigner, E. On an algebraic generalization of the quantum mechanical formalism. *Annals of Math.* (2) **35** (1934), 29–64, particularly pp. 51–54.

[18] Lee, H.-C. Sur le théorème de Hurwitz–Radon pour la composition des formes quadratiques. *Comment. Math. Helv.* **21** (1948), 261–269.

[19] Manning, H. P. *Geometry of four dimensions* (New York, 1914; Dover reprint, 1956).

[20] Radon, J. Lineare Scharen orthogonaler Matrizen. *Abh. Math. Sem. Univ. Hamburg* **1** (1922), 1–14.

[21] Semple, J. G. On complete quadrics (II). *J. London Math. Soc.* **27** (1952), 280–287.

[22] Serre, J.-P. *Représentations linéaires des groupes finis* (Paris, 1967).

[23] Severi, F. Sulla varietà che rappresenta gli spazi subordinati di data dimensione immersi in uno spazio lineare. *Annali di Mat.* (3) **24** (1915), 89–120. Reprinted in *Memorie Scelte*, 405–440.

[24] Stiefel, E. Richtungsfelder und Fernparallelismus in n-dimensionalen Mannigfaltigkeiten. *Comment. Math. Helv.* **8** (1936), 305–353.

[25] Tits, J. Sur la trialité et certains groupes qui s'en déduisent. *Publ. Math. I.H.E.S.* **2** (1959), 13–60.

[26] Todd, J. A. Conics in space, and their representation by points in space of nineteen dimensions. *Proc. London Math. Soc.* (2) **36** (1932), 172–206.

[27] Wong, Y.-C. *Isoclinic n-planes in Euclidean 2n-space, Clifford parallels in elliptic (2n-1)-space, and the Hurwitz matrix equations* (American Math. Soc. Memoirs, No. 41, 1961).

INDEX

Common Sense, the Turing Test,
and the Quest for Real AI

Common Sense, the Turing Test, and the Quest for Real AI

Hector J. Levesque

The MIT Press
Cambridge, Massachusetts
London, England

First MIT Press paperback edition, 2018

© 2017 Massachusetts Institute of Technology

All rights reserved. No part of this book may be reproduced in any form by any electronic or mechanical means (including photocopying, recording, or information storage and retrieval) without permission in writing from the publisher.

This book was set in Stone Sans and Stone Serif by Toppan Best-set Premedia Limited.

Library of Congress Cataloging-in-Publication Data

Names: Levesque, Hector J., 1951- author.
Title: Common sense, the Turing test, and the quest for real AI / Hector J. Levesque.
Description: Cambridge, MA : MIT Press, [2017] | Includes bibliographical references and index.
Identifiers: LCCN 2016033671 | ISBN 9780262036047 (hardcover : alk. paper)—9780262535205 (pb)
Subjects: LCSH: Thought and thinking. | Intellect. | Computational intelligence. | Artificial intelligence--Philosophy.
Classification: LCC BF441 .L483 2017 | DDC 006.301--dc23 LC record available at https://lccn.loc.gov/2016033671

For Pat

Contents

Preface

This is a book about the mind from the standpoint of artificial intelligence (AI). But even for someone interested in the workings of the human mind, this might seem a bit odd. What do we expect AI to tell us? AI is part of computer science after all, and so deals with *computers*, whereas the mind is something that *people* have. Sure, we may want to talk about "computer minds" someday, just as we sometimes talk about "animal minds." But isn't expecting AI to tell us about the human mind somewhat of a category mismatch, like expecting astronomy to tell us about tooth decay?

The answer is that there is not really a mismatch because computer science is not really that much about computers. What computer science is mostly about is *computation*, a certain kind of process, such as sorting a list of numbers, compressing an audio file, or removing red-eye from a digital picture. The process is typically carried out by an electronic computer of course, but it might also be carried out by a person or by a mechanical device of some sort.

The hypothesis underlying AI—or at least one part of AI—is that ordinary *thinking*, the kind that people engage in every day, is also a computational process, and one that can be studied without too much regard for who or what is doing the thinking.

It is this hypothesis that is the subject matter of this book.

What makes the story controversial—and more interesting, perhaps—is that there is really not just one kind of AI these days, nor just one hypothesis under investigation. Parts of AI do indeed study thinking in a variety of forms, but other parts of AI are quite content to leave it out of the picture.

To get an idea of why, consider for a moment the act of balancing on one leg. How much do we expect thinking or planning or problem-solving to be involved in this? Does a person have to be knowledgeable to do it well? Do we suppose it would help to read up on the subject beforehand, like *Balancing on One Leg for Dummies*, say? An AI researcher might want to build a robot that is agile enough to be able to stand on one leg (for whatever reason) without feeling there is much to be learned from those parts of AI that deal with thinking.

In fact, there are many different kinds of AI research, and much of it is quite different from the original work on thinking (and planning and problem-solving) that began in the 1950s. Fundamental assumptions about the direction and goals of the field have shifted. This is especially evident in some of the recent work on machine learning. As we will see, from a pure technology point of view, this work has been incredibly successful, more so, perhaps, than any other part of AI. And while this machine learning work leans heavily on a certain kind of statistics, it is quite different in nature from the original work in AI.

One of the goals of this book is to go back and reconsider that original conception of AI, what is now sometimes called "good old-fashioned AI," and explain why, even after sixty years, we still have a lot to learn from it—assuming, that is, that we are still interested in exploring the workings of the mind, and not just in building useful pieces of technology.

A book about AI can be many things. It can be a textbook for an AI course, or a survey of recent AI technology, or a history of AI, or even a review of how AI has been depicted in the movies. This book is none of these. It is a book about the ideas and assumptions behind AI, its intellectual underpinnings. It is about why AI looks at things the way it does and what it aspires to tell us about the mind and the intelligent behavior that a mind can produce.

It is somewhat disheartening to see how small this book has turned out to be. The entire text will end up taking about 150KB of storage. For comparison, just one second of the Vivaldi concert on my laptop takes more than twice that. So in terms of sheer raw data, my laptop gives equal weight to this entire book and to a half-second of the video, barely enough time for the conductor to lift his baton.

This can be thought as confirming the old adage that a picture—or a single frame of a video—is worth a thousand words. But I think there is another lesson to learn. It also shows that compared to pictures, words are amazingly compact carriers of meaning. A few hundred words might be worth only a momentary flash of color in a video, but they can tell us how to prepare boeuf bourguignon. The human animal has evolved to be able to make extremely good use of these ultra-compact carriers of meaning. We can whisper something in somebody's ear and be confident that it can have an impact on their behavior days later.

How the mind is able to do this is precisely the question we now set out to explore. My hope is that the reader will enjoy my thoughts on the subject and get to feel some of the excitement that I still feel about these truly remarkable ideas.

Acknowledgments

I made my living writing technical papers about AI, more or less. So you might think that writing something nontechnical like this book would be a piece of cake, a stroll in the park. Uh, no. This book was certainly a labor of love, but it was harder for me to write than anything technical.

I was fortunate to have a number of people help me with it. Ernie Davis encouraged me to get started in the first place, and was kind enough to read and comment on a rough first draft. I was also glad to get comments on a subsequent draft from Toryn Klassen, my brothers John and Paul, my daughter Michelle, and four anonymous reviewers. They helped to at least steer the book in a direction, since apparently I had somehow managed to leave port without much of an idea of where I was headed. I'm also grateful for comments on a later draft from Gerhard Lakemeyer, Don Perlis, Vaishak Belle, and Gary Marcus. The book is much better thanks to all their efforts. But of course, they are not to blame for any errors and confusions that remain.

I also want to thank the people at the MIT Press. Marie Lufkin Lee was kind and encouraging even when I had nothing more than a sketchy draft. Kathleen Hensley managed a lot of the practical details, and Michael Sims handled the copyediting, transforming my shaky grammar and wording into something more sturdy.

I wrote some of the book while I was a visiting researcher in Rome. I want to thank Giuseppe de Giacomo and the good folks at the Sapienza—Università di Roma for their friendship and hospitality. I actually completed the bulk of the work after I had retired from the University of Toronto. To have the chance to put as little or as much time as you want on a project of your own devising, how good is that?

I would also like to thank my family and friends who cheered me on from the start. "Why don't you try writing something that normal people can read?" they would sometimes ask, or more diplomatic words to that effect. This book might be the closest I ever get, even if it remains a tad academic in spots. But in case I don't get a chance to do anything like it again, I want to express my gratitude more broadly.

Let me start by thanking my brother Paul who drove me to Toronto to see the movie *2001: A Space Odyssey* when it first came out. As a high-school student, I'm not sure I "got it" at the time, but I did get the impact, the incredible attention to detail. While others were marveling at the trippy aspects of the movie (it being the late sixties and all), I was marveling at the idea of a computer like HAL. It changed my life, and you will see references to *2001* throughout this book.

I began as a university student soon thereafter, thanks to the efforts of a high-school guidance counselor who got me a scholarship in engineering. But I found engineering incredibly tedious, demanding but not rewarding, except for a single course on computers. I knew at that point that Marvin Minsky, who had consulted on *2001*, was on the faculty in engineering at MIT. So I wrote him a letter asking about this. I want to thank the late Professor Minsky for taking the time and trouble to write to a first-year undergraduate—this was well before the days of email!—encouraging him to get out of electrical engineering and into computer science. I might not have been able to walk away from my engineering scholarship had my grandfather not left money for his grandchildren's education. But he did, and so I walked, and this too changed my life.

In computer science, I was very fortunate to come under the wing of John Mylopoulos, the one faculty member in Toronto interested in AI at the time. He taught a new course on AI that I took and loved, he gave me a summer job, he took me on as a graduate student, and he helped me very patiently as I meandered toward a PhD topic some distance away from his area of interest. He also introduced me to Ron Brachman, who was visiting from Boston. Ron was kind enough to offer me a summer job in Boston, and after my PhD, a job in California, where I got to work with him for three glorious years. John also worked very hard to get me an academic position back in Toronto, which ended up being a dream job for me for the rest of my career. John and Ron are the only two bosses I have ever known, and I owe my career to the two of them.

I can hear the music in the background getting louder, so let me wrap this up by thanking my dear wife Pat for having been there for me throughout my career, through ups and downs, from undergrad to pensioner. Like it says on the first page, this one is for her.

Toronto, July 2016

1 What Kind of AI?

This is a book about the mind from the standpoint of artificial intelligence (AI). We will have a lot to say about the mind, but first, we turn to the area of AI itself and briefly introduce the ideas that will be covered in more depth in the rest of the book.

Adaptive machine learning

When it comes to technology, AI is all the rage these days. Hardly a week goes by without some mention of AI in the technology and business sections of newspapers and magazines. The big computer companies like Microsoft, IBM, Google, and Apple are all heavily invested in AI research and development, we are told, and many others are following suit. In November 2015, Toyota announced a $1B investment in AI, and in December 2015, Elon Musk announced a new nonprofit company called OpenAI with yet another $1B of funding.

What is all this buzz about anyway? If you look in more detail at what those billions of dollars are expected to lead to, the technology appears to be quite different from the sort of AI previously imagined in science-fiction books and movies. There is very little talk of intelligent humanoid robots (like those seen in the movie *Blade Runner*, say) or even of a high-powered disembodied intelligence (like the HAL 9000 computer in the movie *2001: A Space Odyssey*).

The kind of AI being contemplated by the technology companies today might better be called *adaptive machine learning*, or AML. The idea in AML, very generally speaking, is to get a computer system to learn some intelligent behavior by training it on massive amounts of data. Indeed, the current excitement about AI leans heavily on the potential application of what is usually referred to as "big data."

This is not the place to try to explain in any great detail how AML technology is supposed to work, but here is the rough idea.

Imagine, for the sake of argument, that we are interested in building a computer system that can recognize cats. Its job, in other words, will be to classify pictures it is given into two groups: those that depict cats and those that do not. The question then is how to build such a system. In the past, AI programmers would have tried to write a program that looked for certain specific cat features in an image. It might look for a cat face: greenish or yellowish eyes with almond-shaped vertical pupils, an inverted pinkish triangle making up the nose, whiskers, and so on. Or perhaps it might look for a cat outline: a small head, triangular cat ears, four legs, and a tail that often points straight up. Or it might look for the distinctive coloring of cat fur. If the program spots enough of these features in an image, it would label the image as one of a cat, and otherwise reject it.

But AML suggests a very different way of doing the job. You start by giving your system a very large number of digitized images to look at, some of which have cats in them and some of which do not. Then you tell your system to *compress* all this visual data in a certain way, that is, to look for a set of "features" that are seen in many patches of many images. A feature might be an area of a certain uniform color and brightness. It might be an area where there is a distinctive change in brightness and color corresponding to an edge. The idea is to find a set of features such that it is possible to reconstruct something similar to the original images as combinations of those features. Then you tell your system to abstract from those features, and to look for common features of those features. And so on for a few more levels.

All this is done without telling the system what to look for. If the original images had many cats in them, there is a good chance that the system will end up with some high-level cat-like features. What's important is that these features are determined by the images themselves, not by what an AI programmer thought was important about recognizing a cat in them. As Andrew Ng from Stanford puts it, "You throw a ton of data at the algorithm and you let the data speak and have the software automatically learn from the data."

This "unsupervised" approach to AML has been found to work extremely well, beyond anything AI researchers might have predicted even just a few decades ago. The success of AML is usually attributed to three things: truly massive amounts of data to learn from (online, in specialized repositories, or from sensors), powerful computational techniques for dealing with this data, and very fast computers. None of these were available even thirty years ago.

Of course, nobody wants to spend billions of dollars on cat recognition. But imagine that instead of cats, the pictures are mammograms, some of which contain tumors that may be hard for doctors to see. Or imagine that the data is not visual at all. It might be made up of sound recordings, some of which contain words spoken by a person of interest. Or imagine that the data involves banking transactions, some of which involve fraud or money laundering. Or maybe the data involves groups of online shoppers characterized by their browsing and purchase history. Or maybe the data is automobile pedal and steering-wheel activity in response to visual data from the windshield. Systems that can learn automatically from massive amounts of data might be applied in a variety of areas of great economic and social significance.

Good old-fashioned AI

Exciting as this AML technology may turn out to be, it is *not* what this book is about. In fact, as will become clear, we will have very little to say about AI technology of any kind, except at the very end, in chapter 10.

What we will mostly be concerned with in this book is a rather different vision of AI first proposed by John McCarthy, Marvin Minsky, Allen Newell, Herb Simon, and others in the 1950s, something we call *Good Old-Fashioned AI*, or GOFAI for short. These researchers emphasized a different sort of intelligence, one based not on learning from massive amounts of data, but on *common sense*.

What is common sense? How is having common sense any different from being well trained on large amounts of data? This will take much of the book to sort out. One difficulty in trying to be clear about the difference is that in our normal everyday lives, we spend almost all of our time in situations that are at least somewhat familiar to us. We rely to a very large extent on our routines, patterns of behavior that we have learned over time. I usually pick up a cup of coffee on my way to work. I usually greet my coworker George when I see him in the hall. I usually put mustard on hot dogs. Even if I happen to see a tree I have never seen before, it is still close enough to what I have seen for it to feel quite familiar to me. There is certainly no surprise or disorientation.

However, as we will see in much more detail in chapter 7, common sense enters the picture when we are forced to act in situations that are sufficiently unlike the patterns we have seen before. What exactly do I do on that one morning when the door to my coffee shop does not open? Do I just keep tugging at it? What do I say to my colleague George if the first thing he says to me is "Did you hear about the nasty chemical spill on the third floor?" Do I just say "Good morning" as usual and move on? What do I do if I am given a hot dog that is already fully loaded with condiments? Put more mustard over the whole mess? It is how we behave in new, unfamiliar situations, when we say to ourselves "Hold on a minute, what's going on here?" that really shows whether or not we have common sense.

The question that motivates the researchers in GOFAI is this: What accounts for this common sense? How do people use it to figure out what to do when their usual routines no longer apply? What is this "figuring out" anyway? What goes into a decision about what to do with a door that won't open? When I decide to step outside of a building without a coat on a cold winter's day just because of something I was told, how does that work? How is that one person making noises with his mouth—saying certain words—can be enough to cause me to behave in a way that is totally unlike my normal routine?

This is the mystery that early AI researchers were intrigued by and *this* is what this book is about.

Programs with common sense

The early pioneers in AI met at a conference at Dartmouth College in the summer of 1956. This is where AI was first given its name by the organizer of that conference, the American computer scientist John McCarthy (1927–2011). A direction for research was laid out in a brilliant and quite unprecedented paper he wrote in 1958 entitled "Programs with Common Sense."

We will have more to say about the technical aspects of this paper in chapter 9, but one thing it made explicit was a focus on common sense. McCarthy had something very definite in mind:

We shall therefore say that a program has common sense if it automatically deduces for itself a sufficiently wide class of immediate consequences of anything it is told and what it already knows.

As we will see again in chapter 3, McCarthy was especially intrigued by how people use what they know and what they have been told to figure out what to do.

However, even from the very start, common sense was not considered to be something distinct from learning. McCarthy also says: "Our ultimate objective is to make programs that learn from their experience as effectively as humans do."

Clearly we do learn something by trying to open a door that happens to be locked. We learn from our coworker George about a chemical spill. We learn by direct observation that a hot dog is fully loaded.

Yet the emphasis in GOFAI is different from AML. As we will see in chapter 5, learning in GOFAI need not involve large amounts of training data, and it often deals with *language*. John McCarthy again: "In order for a program to be capable of learning something it must first be capable of being told it." The primary focus is very clearly on the linguistic aspects of learning, and he called his proposed system the "advice taker."

Indeed, GOFAI is sometimes criticized as being overly preoccupied with symbols and words. Intelligent behavior, the critics rightly point out, need not be something you can put into words. You might be able to ride a motorcycle quite well without being able to say in words how far over you should lean on a tight curve. You might be able to pick out two brothers in a group of boys without being able to articulate what it is that allows you to spot that family resemblance.

And yet, it is true that human language holds a position of honor in GOFAI. And there are two very good reasons for this. First, as will be argued in chapter 6, quite apart from communication, language plays a very special role in human behavior, a role not seen in other animals: much of how we deal with new situations involves using what we have read or been told earlier using language.

Second, language is a superb medium for exploring intelligent behavior. If we want to find out how a person would deal with a new and unfamiliar situation, we can ask him or her. While it is possible to exhibit intelligence without saying a thing, it certainly requires intelligence to be able to use language the way we do, and so it is a supreme test of intelligence, something first suggested by Alan Turing.

The Turing Test

Alan Turing (1912–1954) was a British mathematician involved in the early development of computers. (We will have more to say about him and his role in chapter 8.) He was one of the first to consider seriously whether computers could be programmed to do tasks that required intelligence in people, such as playing a reasonable game of chess. While it is true that computers can do only what they are programmed to do, it was far from clear then—and it is still far from clear today—just what we could program them to do.

I suspect that Turing grew tired of philosophical discussions about whether a machine would ever be able to think, or understand something, or have consciousness. He foresaw some of the incredible technical challenges posed by AI, but was perhaps exasperated by arguments to the effect that even if all these challenges could be overcome, the resulting computer system would never be said to think or understand or be conscious for other reasons he had no control over, such as the fact that the computers were not biological.

In a very influential 1950 paper, Turing suggested that we should shift our attention away from how the machine was made, what it looked like, or what may or may not be going on inside it, and concentrate instead on its *externally observable behavior*. Of course, there are different sorts of behavior that we might want to consider. We might put the machine in some new and unfamiliar situation and see what it does. We might ask it to recognize pictures of cats. We might study what it does with a fully loaded hot dog. What Turing came up with was the idea of participating in a no-holds-barred conversation with an interrogator.

This is what he called the Imitation Game (described in the 2014 movie of the same name). The idea is that there would be an extended conversation over a teletype between an interrogator and two hidden participants, a person and a computer. The conversation would be natural, free-flowing, and about any topic whatsoever. The computer would win the game if no matter how long the conversation, the interrogator could not tell which of the two participants was the person. In current parlance, the computer is said to have *passed the Turing Test.*

It is important to remember that as far as Turing was concerned, the conversation could be about anything. Here is a typical conversation imagined in his paper:

Interrogator: In the first line of your sonnet which reads "Shall I compare thee to a summer's day," would not "a spring day" do as well or better?
Computer: It wouldn't scan.
Interrogator: How about "a winter's day?" That would scan all right.
Computer: Yes, but nobody wants to be compared to a winter's day.
Interrogator: Would you say Mr. Pickwick reminded you of Christmas?
Computer: In a way.
Interrogator: Yet Christmas is a winter's day, and I do not think Mr. Pickwick would mind the comparison.
Computer: I don't think you're serious. By a winter's day one means a typical winter's day, rather than a special one like Christmas.

We are, of course, very far away from a computer program that can carry on a conversation like this. But is there some reason to believe that we will never be able to write such a program? *This* is what Turing suggests we should focus on.

Turing's point in all this, it seems to me, was the following: These mental terms like "intelligent," "thinking," "understanding," and so on are really too vague and emotionally charged to be worth arguing about. If we insist on using them in a scientific context at all, we should be willing to say that a computer that can pass a suitable behavioral test has the property in question as much as a person, and leave it at that. Adapting the dictum of the title character in the movie *Forest Gump* who said "Stupid is as stupid does," we can imagine Turing saying "Intelligent is as intelligent does." In other words, instead of asking "Can a machine think?" we should really be asking "Can a machine behave like a thinking person?"

The Chinese Room

Turing's proposal is perhaps an obvious one, at least in retrospect. His emphasis on externally observable behavior is really at the heart of all AI research, as we will see in the next chapter. (We shall have reason to question the use of an *informal conversation* in a test for intelligence, but we will get to that in chapter 4.) Nonetheless there have been detractors. We will briefly discuss one objection to Turing's proposal that came out of the philosophical literature. (Readers less interested in philosophical wrangling should jump ahead to the next chapter.)

In the 1980s, the philosopher John Searle argued that there was much more to understanding (or thinking, or being intelligent) than merely getting some observable behavior right, even if that behavior is as general and versatile as the Imitation Game. Here is (a slight variant of) his argument.

Imagine that there already exists a computer program that is somehow able to pass the Turing Test, that is, that can engage in an unrestricted conversation as well as a person. Let us suppose that this conversation takes place in Chinese, not in English. So in accordance with the Imitation Game, Chinese characters are presented to this program (encoded in some way) as input, and the program is able to produce Chinese responses (also encoded in some way) as output that cannot be distinguished even over a long period of time from those of a person who really understands Chinese.

Let us now imagine that Searle himself does not know Chinese, but knows computer programming very well. (In reality, he does not, but no matter.) He is hidden in a room with a book that contains the complete text of the computer program. When someone outside the room slips him a piece of paper with Chinese written on it, he does not understand what it says, but he can simulate what the computer program would do with it. He traces the behavior of the program using the book, writes on a piece of paper the output that the program would produce, and hands that back, again without understanding what any of it means.

So what do we have here? There is a person, John Searle, who is receiving Chinese messages in his room and using the book to produce Chinese responses that are indistinguishable from those of a native Chinese speaker (since the computer program is assumed to pass the Turing Test). His externally observable behavior, in other words, is *perfect*. And yet Searle does not understand Chinese! Searle's conclusion: getting the behavior right is not enough, and so Turing is wrong.

An objection to this argument is that it is not Searle who is producing this behavior, but Searle together with the book containing the text of the computer program. Although Searle does not understand Chinese, the overall *system* consisting of Searle and the book does. So Turing is not wrong. Searle's reply to this objection is beautiful in its simplicity: imagine that he *memorizes* the book and then destroys it. Then there is no longer a system to talk about; there is just Searle. So Turing *is* wrong.

This is the Chinese Room argument, in a nutshell.

What is there to say about it? Understandably, there is impatience with it in the AI community. There is much more interest in AI in the technical challenges involved in producing such a computer program (assuming it is possible at all), than in what conclusions we would draw about a person in a room who can accurately simulate the running of such a program (maybe millions of times more slowly).

But in terms of the thought experiment itself, here is one thing to think about: how can we be so sure that Searle does not come to *learn Chinese* by memorizing the book? If he ended up knowing Chinese afterward, the argument about passing the Turing Test without understanding Chinese would no longer apply. So we have to ask: why should we believe that there can be a book that does the job Searle wants *and* that can be memorized without learning to speak Chinese?

It's tough to answer this question one way or another without a better idea of what the computer program in the book would have to be like. One argument I have made in the past involves turning to a different and much simpler form of behavior, the summation of numbers.

Here's my idea. Imagine that instead of speaking Chinese, the behavior under examination involves the ability to add twenty ten-digit numbers, no more, no less. A book listing every possible combination of twenty ten-digit numbers and their sums would allow a person who does not know how to add to get the summation behavior right. Any time the person is asked what a sum is, the correct answer could be found by looking it up in the book, just like Searle suggested for Chinese.

The main thing to observe about such a book is that it cannot exist. It would need to contain 10^{200} distinct entries for all the combinations of numbers, but our entire physical universe only has about 10^{100} atoms.

There is another smaller book that would do the job, however, which is an English language description of how to add:

first, a 10 × 10 table for the addition of single digits, then the right-to-left process of addition for two numbers (with carry digits), and finally, the process of multinumber addition. This second book can definitely exist; it would only be a few pages. What is especially interesting about this second book is that a person who does not know how to add but who memorizes the instructions in the book would thereby learn how to add!

This is enough to cast doubt on Searle's Chinese Room argument. If there is no book that would allow a person to get something as simple as the summation behavior right without also teaching the person how to add, or at least how to add twenty ten-digit numbers, why should we think that conversing in Chinese will be any different?

However, this doesn't refute Searle's argument either. In the end, we really don't know what it would be like to memorize Searle's book because we don't know what a computer program for Chinese would need to be like. And the only way to find out is to tackle those technical challenges, just as Turing suggested.

2 The Big Puzzle

Intelligent behavior in people is a product of the mind. But the mind itself, of course is not a thing; it is more like what the human brain *does*. The actual physical object that lies behind our external behavior is the brain.

The problem

Although we know much more about the human brain than we did even ten years ago, the thinking it engages in remains pretty much a total mystery. It is like a big jigsaw puzzle where we can see many of the pieces, but cannot yet put them together. There is so much about us that we do not understand at all. Just how are we different from the rest of nature? What makes us so darn smart?

People in different areas of research come to this puzzle from different angles, each holding some of the pieces. Not too surprisingly perhaps, each of them argues that *their* perspective is the important one, the one that really matters. Here is the sort of thing you might expect to hear:

• What we need to care about is *language*, since the presence of a mind is most clearly manifest in the ideas it expresses. We know a mind by listening to what it says, reading what it writes. We are unique in the animal kingdom in recording what we know, what we have learned, how we think. We will never get to study the brain of Shakespeare, but we get to experience the mind of Shakespeare every time we tackle one of his plays.

• What we need to care about is *psychology*, how a mind works in people and how that compares and contrasts with other animals. People's intuitions about how they think is quite different from how they really do think when studied closely in a controlled setting. Memory, learning, perception, attention, cognition, affect, consciousness, these are all psychological categories and together make up the mental life of people.

• What we need to care about is *neuroscience*, how a brain organizes itself to make a mind. When we talk about the mind, what we are really talking about are the states and transitions of a large assembly of highly connected electrochemical components. As we learn more and more about the brain—and the recent progress here has been phenomenal—there will be a revolution in how we talk about the mind, just as there has been a revolution in how we talk about life since the discovery of DNA.

• What we need to care about is *evolution*, how the human species developed under evolutionary pressure. Nothing in the human species, including its brain, stands alone. Every human faculty is the product of evolution, and correlates can always be found in other species. The human mind is a Swiss army knife of adaptations, and one cannot understand it without understanding how it evolved and how it connects to reproductive success.

And there are others.

I believe a certain amount of humility is called for when we talk about things like the brain, the mind, thinking, and intelligence. Listening to scientists brimming with confidence in full promotion mode, it may appear that new ideas, new approaches, new techniques, new tools are putting us on the threshold of an era where we will crack the mystery. I think this is quite wrong. The mistake, in fact, is to think that there is a single mystery to crack. What we have is a *collection* of mysteries, a jigsaw puzzle of pieces that need to be brought together and assembled from many different perspectives before we can truly appreciate the overall picture.

We need to train ourselves to be skeptical of any research group that insists that one part of this puzzle is the true core, the key to the whole thing. We should be even more skeptical of anyone who claims something like "We are on the verge of figuring out how the brain works." This should sound to us like someone claiming to have figured out how weather works, or how the stock market works.

Just for reference, let us call the issue of confusing a handful of pieces for the entire puzzle the *Big Puzzle* issue. (It will come up again often.)

Given this Big Puzzle issue, how are we then to talk about thinking at all without falling into pompous oversimplifications? I believe the answer is to accept that we are dealing with only one part of the puzzle, and to do what we can to see how the pieces we are holding fit together, while resisting the temptation to suggest that the rest of the puzzle is just more of the same. The mind is staggeringly complex, but this should not rule out looking at a small part of it in detail.

So what part of the puzzle will be discussed in this book? To jump the gun somewhat, this is the story that will be told here:

• What we need to care about is *intelligent behavior*, an agent making intelligent choices about what to do. These choices are made intelligent through the use of background information that is not present in the environment the agent is acting in. This background information is what we call *knowledge*. The application of knowledge to the behavior at hand is what we call *thinking*. The problem that needs to be sorted out is how, in concrete terms, this all works, how background knowledge can make a difference in an agent deciding what to do. The solution we consider is a *computational* one. In the same way that digital computers perform calculations on symbolic representations of numbers, we suggest that human brains perform calculations on symbolic representations of knowledge, and then use the result of those calculations to decide how to act.

It should be clear that this story is very far from proposing a solution to the entire puzzle. If it were, where would emotions like modesty, envy, and grief fit in? What about the social or interactional aspects of intelligence? Where is perception and imagination? Or daydreams and fantasies? What parts are uniquely human? What accounts for degrees of intelligence? And what about different kinds of intelligence (like emotional intelligence)? Or the effect of drugs and mental illness? And where is consciousness in all this? Or creativity? And what about spirit and pluck, the stuff that makes Molly Brown unsinkable?

These are all great questions that I will not be talking much about. Yet I do not want to give the impression that what is left is small or beside the point. Thinking, even in the very narrow sense described above, is still as broad as the ideas that can be thought about. And we can think about anything! Despite all the amazing scientific progress to date, we still have no clear sense of what happens when we think about who will win the Academy Award for Best Actor, or how thinking about Debussy is similar to but different from thinking about Ravel, or what it is like to change your mind over a period of time about whether a free market needs to be regulated.

The mystery remains even in the small fragment of the puzzle considered here.

The difficulty

Before leaving this topic, it might be worth looking in more detail at why areas like psychology or neuroscience by themselves are unlikely to be able to tell us what we want to know about how the mind works. To make the point, we will use a thought experiment involving a simple machine that will be much easier to analyze.

Imagine a small device connected to a beeper, a lightbulb, and a keyboard. Anytime someone types a two-digit number on the keyboard, the machine responds with some beeps and some flashes. Call the machine M. Think of M as a very simple brain, with the keyboard as its only sense organ, and the beeper and lightbulb as its only effectors. Imagine that your job is to figure out why it is beeping and flashing the way it does. The exercise is useful because it gives us a good feel for what it might be like to figure out what is behind intelligent behavior.

Let us say, for example, that the numbers 37, 42, 53, 16, and 37 are typed in sequence on the keyboard. M produces the behavior shown in the following table:

Input	Beeps	Flashes
37	1	6
42	1	8
53	2	8
16	2	2
37	3	2

Now given all this, what makes it behave the way it does?

I am going to start by giving away the answer. M is a small digital computer attached to the keyboard, beeper, and light-bulb. It repeatedly takes as input the number typed on the keyboard, and produces as output a number of beeps and of flashes according to a tiny computer program.

The entire program that controls M is displayed in figure 2.1. (Readers who do not wish to try to read this small computer program can also skip the rest of this paragraph.) To see how it works, note that it uses integer arithmetic, where *mod* means the remainder after division. So the value of 37 / 10 is taken here to be 3, and the value of 37 mod 10 is 7. For example, when the 37 is first typed on the keyboard, M produces 1 beep and 6 flashes. Here's why: according to the program in figure 2.1, the W is set to 37, the X is set to 3, the Y is set to $3 \times 3 + 7 = 16$, where 16 / 10 = 1 and 16 mod 10 = 6. When 37 is typed on the keyboard the second time (the fifth input), the result is 3 beeps and 2 flashes because the U at that point is 5 (from the third input), and since $5 > 3$, the Y is set to $5 \times 5 + 7 = 32$.

So that's the secret. Now having seen it, pretend that you know nothing about the program in figure 2.1, but that your job is to understand why M does what it does.

We can imagine being a psychologist, running some experiments on M and observing its behavior. There are fewer than a

```
integer U, V, W, X, Y

set U to 0

set V to 0

repeat the following forever:

    set W to the next typed number

    set X to W / 10

    if X > U

        then set Y to X*X + (W mod 10)

        else set Y to U*U + (W mod 10)

    beep (Y / 10) times

    flash (Y mod 10) times

    set U to V

    set V to X

end repeat
```

Figure 2.1
The code for a mystery machine

hundred possible inputs here, but even in this very simple set-ting, our life is complicated by the fact that *M* has memory and makes decisions about how to behave based not just on the last number it saw.

To get an idea of what it is like for a real psychologist, we have to imagine that the number of tests we can run is much, much smaller than the range of possible inputs available to the machine. Think of a reading comprehension test (like the kind to be discussed in chapter 4), and compare the number of sen-tences in the test to the number of sentences that subjects will read in their entire lives. Psychology must live with evidence drawn from very small parts of the total behavior space.

For example, let us simply generalize M so that instead of taking as input a two-digit number, it takes as input a ten-digit number. Now instead of 10^2 (one hundred) possible inputs, we have 10^{10} (ten billion) of them. If we observe that, because of memory, we may also have to consider what the machine saw in the previous step and the one before that, then we need to consider 10^{30} possible sequences. If the memory could go back further and depend on the ten most recent inputs, there are now 10^{100} sequences to sample from, more than there are atoms in the known universe.

So it does not take much to rule out the possibility of observing a significant range of stimulus and response. The sensory environment of M, simple as it is, and its memory, simple as it is, overwhelm any sort of comprehensive testing.

This, in a nutshell, is what makes psychology so hard.

It is extremely difficult to design an experiment constrained enough to provide meaningful results. If my subjects are Jim and Jane, how am I going to control for the fact that they have had very different lives, seen very different things, and come to my experiment with very different beliefs and goals? Not too surprisingly, some of the most revealing psychological experiments involve perceptual tasks where an immediate response is required, in the milliseconds, say. This is quick enough that long-term memory, which will be quite different for Jim and for Jane, plays a less significant role. When a subject gets to sit back and muse for a few seconds, it is extremely difficult to control for all the variables.

In a way, psychology is handicapped by the fact that it gets to look at subjects only from the outside. External stimuli can be presented, and external responses can be observed, but that's it. It is not ethical to open up a living person's head and attach electrodes here and there to see what is happening. Some of our best understanding of brain function has come from cases where the brain was exposed as a result of an operation—severing the *corpus callosum* to control epilepsy, for example—and asking patients to describe what they sense when parts of the brain are stimulated.

However, there is new technology, such as fMRI, that is much less invasive than brain surgery and is giving us a much better picture of many parts of brain function. We can see that the part of the brain involved in motor control is activated even when a subject is just thinking about physical activities. We can see that the part of the brain that is active when a person is swearing is not the one that is normally active during polite language production. All these very impressive developments in neuroscience suggest that given sufficient time, we will understand how beliefs and goals come together in thinking to produce human behavior.

To see why we should nevertheless be skeptical about this, let us go back to M. Imagine a neuroscientist whose job it is to determine why M behaves the way it does. Like the psychologist, the neuroscientist is not told about the program in figure 2.1. Unlike the psychologist, however, the neuroscientist will be given access to the internal workings of M, as if it were a functioning brain.

When M is opened up and studied in the lab, the neuroscientist can see that it is an assembly of standard electrical components powered by a battery. As digits are typed, some of these components are activated and others stay dormant, perhaps only rarely lighting up. As more and more numbers are typed on the keyboard, some tantalizing patterns begin to emerge. The question is: do we expect the neuroscientist to crack the puzzle about M's behavior?

It is certainly true that M is nothing more than a collection of electrical components. Any behavior it produces is ultimately due to those components being in one state or another. If M were a brain we would say that the state of the brain is what determines how we behave; anything else we might choose to talk about (beliefs, goals, emotions, a mind, whatever) must be realized in one way or another in the state of the brain.

But the question is whether we will be able to recover the *regularities* in M's behavior in the properties of M's components. For example, we might want to be able to discover things like the fact that M squares the *first* digit in the number it sees but not the *second*. Will we see this in the electrical components?

There is good reason to believe that we will not. Let us suppose that the neuroscientist is talented and lucky enough to extract the entire program that M is running by carefully studying the state of all the electrical components over a period of time.

Here's the problem. M squares the first digit (or a previous first digit) because of lines 8 and 9 of the program in figure 2.1. But *that* program may not be located anywhere in M's memory. It is typical of computers that programs are first translated into another form that is more easily executed by hardware. In computer jargon, the program in figure 2.1 is called the *source code*, and the translated version that is stored in the memory of M is called the *object code*. At the very best, the talented neuroscientist gets only the object code, and there is no reason to suppose that having the object code in hand would allow someone to recover the source code that generated it.

For example, squaring a number on a computer is typically not a single operation. Multiplication may show up as a large number of operations in object code. (There are clever forms of multiplication that perform much better than the digit-by-digit form we were taught in grade school.)

Even worse, the numbers themselves may not be encoded in a simple way in the states of the electrical components. Multiple components may be involved and they need not be physically close to one another. Indeed, in so called *distributed representations* (used in some neural net models of the brain) it may be necessary to look at the state of many electrical components to find the value of any one number represented. Worst of all, in a distributed representation, the state of a single component may be involved in the representation of more than one number.

So while translating from source code to object code is typically easy, going from object code back to source code is like trying to crack a big cryptographic puzzle. Companies that sell software products protect their intellectual property by releasing object code only, confident that this "reverse-engineering" is difficult enough that it cannot be solved in an economically feasible way. (Software products that are "open source" are counterexamples where the source code is made public as well.)

So even with the finest test equipment and electrical probes, the neuroscientist may not be able to recover M's original source program. Having unrestricted access to all of the components in the lab, even when the components themselves are assumed to be well understood, is no guarantee that we can figure out something as simple as why M behaves the way it does.

This, in a nutshell, is what makes neuroscience so hard.

Even if we had total access to the hundred billion or so neurons making up a human brain, and even if we were able to treat
those neurons as idealized, noise-free, digital components, we
might still not be able to figure out why we behave the way we
do. When a neuroscientist has to deal with *real* neurological
components, not electrical ones, where there are a number of
confounding chemical and biological processes going on, the
situation is that much worse. How do we remember certain
kinds of facts? How do we assemble those facts to draw new conclusions? How do we use these conclusions to decide how to
behave? All of these are clearly tremendously more complex
than figuring out why *M* is beeping three times. It is asking too
much of neuroscience, even a wildly more advanced version of
neuroscience, to tell us what we want to know.

We need to look elsewhere.

An approach

When faced with the problem of making sense of a complex
phenomenon—and human thinking is certainly one of them—
we actually have two options: we can study the objects that
produce the phenomenon (that is, human brains), or we can
attempt to study the phenomenon itself more directly.

Consider, for example, the study of *flight* (in the days before
airplanes). One might want to understand how certain animals
like birds and bats are able to fly. One might also want to try to
build machines that are capable of flight. There are two ways to
proceed:

• study flying animals like birds, looking very carefully at their
wings, their feathers, their muscles, and then construct machines
that emulate birds;

- study aerodynamics—how air flows above and below an airfoil, and how this provides lift—by using wind tunnels and varying the shapes of airfoils.

Both kinds of studies lead to insights, but of a different sort. The second strategy is perhaps the more general one: it seeks to discover the principles of flight that apply to anything, including birds.

Thinking is similar. Although we do want to understand human thinking and how it leads to intelligent behavior, this does not mean that there is no choice but to study humans. While there is a lot to be learned by studying the human brain (and other brains too), we can focus on the *thinking process* itself to determine general principles that will apply to brains and to anything else that needs to think.

This is what the philosopher Daniel Dennett calls taking a *design stance*. What we attempt to do is see what would be involved in designing something that is capable of flying or thinking. Instead of getting caught up in the details of how human brains do the job, we shift our attention to the job itself, and ask how it might get done *at all*. The hope is that in so doing, we get to see what is and is not essential in producing the phenomenon of interest.

Of course a design stance will not solve the entire puzzle. It may not tell us much about how the phenomenon can be the end result of an evolutionary process, for instance. And it is not applicable at all to phenomena that cannot be observed. If what we care about is not how birds are able to fly, but whether they have a certain internal sensation that they do not display, then a design stance will be of no help.

So in the end, what we will be concerned with here is *observable* intelligent behavior and how it is produced. We will argue that thinking is needed, of course, but we will not be concerned with what the result of this thinking might *feel like* to the agent doing the thinking.

For many researchers, it is the subjective feeling of conscious thought (sometimes called *qualia*) that is the truly interesting and distinctive feature of the human mind. I take this to be a Big Puzzle issue. The sensation of consciousness is an interesting part of the puzzle, no doubt, but it is not the only one.

3 Knowledge and Behavior

In the mid-twentieth century, digital computers became much more widely available. Impressed as people were by their speed and accuracy, they were sometimes called "electronic brains." Of course the makeup of computers (then and now) has very little in common with the makeup of biological brains. The more plausible question that came out of all this was whether computers were capable of doing some of the things that human brains were capable of doing.

For John McCarthy, what was most striking about the sort of commonsense behavior exhibited by people was how it was conditioned by *knowledge*. In deciding what to do in novel situations, people were able to use what they had been told and what they already knew about the world around them. What he had in mind were examples like the following:

- In planning how to behave, people had to use knowledge about their current situation, the options available to them, and what effect their actions would have.
- In making sense of a visual scene, people had to use what they knew about the appearances of objects to fill in details of the scene that were hidden from view.

• In using language, people had to use the knowledge they had about the topic of conversation and what the words meant.

A very wide range of intelligent behaviors seemed to be unified by the idea that they all depended in similar ways on a store of background knowledge. Once you know that lemons are yellow, for instance, this helps you do a number of things: paint one in a picture, spot one in a grocery store, make a list of yellow fruits, and others. McCarthy was the first to propose that to understand intelligent behavior we needed to study the knowledge that behavior depends on, and how this knowledge might be applied in deciding how to behave.

Before we get back to the Turing Test and to why McCarthy's proposal still matters in the next chapter, we need to slow down and take a closer look at the idea of knowledge itself and what it means to say that intelligent behavior depends on it.

Beyond stimulus and response

In understanding human behavior, there are many occasions where the simplest and best explanation for why we do what we do is in terms of stimulus and response. This is seen quite clearly in other animals, and we exhibit it too, including when we use language. We get our clothing caught on a nail unexpectedly and say "What the heck?" We bump into a person on a crowded train and say "Sorry." We see a child leaning precariously over a rail and say "Watch out!" We hit our thumb with a hammer and say a number of colorful things.

But it would be a Big Puzzle mistake to think that the rest of our language use was just more of the same. Consider how language is used in the following story:

You are with a group of people talking about the movie *2001: A Space Odyssey*. At one point your friend Tim says "The film is just gorgeous on Blu-ray." You say "It's hard to believe it first came out in 1968."

So what do we have here? There is again a stimulus (Tim's comment) and a response (your comment). But now it is much harder to account for how the stimulus elicits the response. What caused you to say the word "1968," for instance?

There is clearly a *gap* here between stimulus and response, and the only plausible explanation about what fills this gap is some *thinking* on your part. You knew something, namely that the movie was released in that year, and you decided that saying so would advance the conversation in an interesting way. Maybe all you really wanted to do was to support Tim, or to connect with your friends, or to be seen as contributing your share of the conversation. That desire does not by itself explain why you said what you did. What accounts for it is applying what you knew about the movie.

Here is what the psychologist Sir Frederic Bartlett says about this process:

The important characteristics of [the] thinking process, as I am proposing to treat it, can now be stated: The process begins when evidence or information is available which is treated as possessing gaps, or as being incomplete. The gaps are then filled up, or that part of the information which is incomplete is completed. This is done by an extension or supplementation of the evidence, which remains in accord with the evidence (or claims to do so), but carries it further by utilizing other sources of information besides those which started the whole process going, and, in many instances, in addition to those which can be directly identified in the external surroundings. (*Thinking*, 1958, p. 75)

It is these "other sources of information" that make all the difference.

The back story: At some point you must have found out that *2001* was released in 1968. Maybe you saw the movie when it first came out, maybe you saw that date watching the movie later, maybe you read about the date somewhere, maybe you saw it on a poster, maybe you heard somebody talking about it. But however you found out, that stimulus caused a change in you. You might not have said anything at the time, but you started believing that *2001* was released in 1968. (We will have more to say about the difference between "knowing" and "believing" below.)

Later, when the movie is being discussed and Tim says his bit, thoughts occur to you very quickly. You recall watching the Blu-ray and how impressed you were, just like Tim. But you don't just agree with Tim or say "Awesome!" You do not want to appear banal or uninformed, and you quickly form a thought you had never consciously entertained before, which is that *2001* looked better than other movies of the same period. But you don't say that either. You happen to know something specific: the period was 1968. And this is what you bring to the table.

The moral of the story is this:

Knowledge, what you know and can bring to bear on what you are doing, is an essential component of human behavior.

This idea is obvious to many of us. But it does appear to be problematic for many others, mainly certain psychologists. They might say:

Why not simply say that there is one stimulus (reading about *2001*, say) that trains you, and a later stimulus (Tim's comment) that causes you to respond according to your training. Why all this mumbo-jumbo about knowledge and belief that we have no way of accounting for scientifically?

The trouble with the bare stimulus-response story is that it misses something very important. As the philosopher Zenon Pylyshyn has stressed in a related context, the actual stimulus you encountered (when reading about *2001*) is neither necessary nor sufficient to account for the response you produced.

First of all, note that what you saw written on a page is only one among very many visual stimuli that will do the trick. The information can be presented in a way that looks nothing at all like what you saw: different paper, different colors, different fonts, different words, even a different language. And there is really nothing visual about it either. You might have heard someone talking about *2001*. But this could have been quite different too: different volume, different pitch, different intonation, different words, different language.

Going further, maybe you did not even see or hear the word "1968." Maybe you heard that *2001* was released the year before Neil Armstrong walked on the moon. Or maybe all you were told is that the Best Picture that year went to the movie *Oliver!*, which might be enough if you knew enough about *Oliver!*. In terms of what the sensory stimulus needs to look or sound like, the range is incredibly wide.

Furthermore, the words "The movie *2001* was released in 1968" themselves might not do the trick. If they come right after the words "Here is something that is definitely false", you would not respond to Tim in the same way. Or maybe you see the words as an item in a long piece entitled "Common misconceptions about the movies of Stanley Kubrick." Or maybe somebody tells you that *2001* was released in 1968, so that you hear those words, no more, no less, but you have good reason to believe that the person saying them is only guessing or is lying. Again your response to Tim would be different.

In the end, what matters about the stimulus, what causes it to do the job, is not what it looks or sounds like at all. There are many sorts of stimuli in context that will work, and many others that will not. What matters is whether or not they are *enough to get you to believe something*, namely, that *2001* was released in 1968.

Moreover, once you are in this state of belief, how you got there is no longer important. Your belief can affect any number of other actions you later do, linguistic ones and nonlinguistic ones, regardless of the original stimulus. If you decide to sort your Blu-ray movies by decade, or if you decide to have a moviefest of *Great Movies From the Sixties,* or if somebody asks you a direct question about Kubrick, all these actions will be affected by this one belief, however it came about.

Intelligent behavior is, in the words of Pylyshyn, *cognitively penetrable*: the decisions you make about what actions to perform is penetrated by what you believe, just as McCarthy first emphasized. If you come to believe that *2001* was released in 1972, for instance, the actions you choose will all change systematically.

Of course not all our actions are affected by what we believe in this way. Involuntary reflexes, for example, are not. Whether we raise our leg if someone taps us on the knee, or whether we blink as something approaches our eye, or even (to a certain extent) what we say when we hit our thumb with a hammer, these actions are selected very quickly and do not seem to depend on the beliefs we have. But sorting our movies by decade is one action that does.

So to summarize: As with all animals, we act in response to stimuli. In some cases, the mapping is direct: we sense something, and we react to it. But for very many others, the mapping is less direct: we sense something, but how we react depends on the beliefs we happen to have.

Knowledge vs. belief

What exactly do we mean by knowledge? The basic idea is simple: knowing something means taking the world to be one way and not another. But let us go over this a bit more carefully.

First, observe that when we say something like "John knows that ...," we fill in the blank with a declarative sentence. So we might say that "John knows that *2001* was released in 1968" or that "John knows that Mary will come to the party." This suggests, among other things, that knowledge is a relation between a knower, like John, and a proposition, that is, the idea expressed by a declarative sentence of English or some other language, like "Mary will come to the party."

And what are propositions then? They are abstract entities, not unlike numbers, but ones that can be true or false, right or wrong. When we say that "John knows that *P*," we can just as well say that "John knows that it is true that *P*." Either way, to say that John knows something is to say that John has formed a judgment of some sort, and has come to realize that the world is one way and not another. In talking about this judgment, we use propositions to classify the two cases, for example, those where Mary will come to the party, and those where she won't.

A similar story can be told about a sentence like "John hopes that Mary will come to the party." or "John worries that Mary will come to the party." The same proposition is involved, but the relationship John has to it is different. Verbs like "knows," "hopes," "regrets," "fears," and "doubts" all denote what philosophers call *propositional attitudes*, relationships between agents and propositions. In each case, what matters about the proposition is its *truth condition*, what it takes for it to be true: if John hopes that Mary will come to the party, then John is hoping that the world is a certain way, as classified by the proposition.

Of course, there are sentences involving knowledge that do not explicitly mention propositions. When we say "John knows who Mary is taking to the party," or "John knows how to get there," we can at least imagine the implicit propositions: "John knows that Mary is taking ... to the party," or "John knows that to get to the party, you go two blocks past Main Street, turn left ...," and so on. On the other hand, when we say that John has a deep understanding of someone or something, as in "John knows Bill well," or a skill, as in "John knows how to skate backwards," it is not so clear that any useful proposition is involved. (While this type of "procedural" knowledge is undoubtedly useful, we will not have much more to say about it, except briefly in the section "Learning behavior" in chapter 5.)

A related notion that we will be very concerned with, however, is the concept of *belief*. The sentence "John believes that *P*" is clearly related to "John knows that *P*." We use the former when we do not wish to claim that John's judgment about the world is necessarily accurate or held for appropriate reasons. We also use it when the judgment John has made is not based on evidence, but is perhaps more a matter of faith. We also use it when we feel that John might not be completely convinced. In fact, we have a full range of propositional attitudes, expressed in English by sentences like "John is absolutely certain that *P*,"

"John is confident that *P*," "John is of the opinion that *P*," "John suspects that *P*," and so on, that differ only in the level of conviction they attribute to John.

Clearly how a person uses what he or she believes depends on how certain the person is about what is believed. This is called the *degree of belief*. For now, it will suit our purposes to not distinguish between knowledge and belief, or to worry about degrees of belief. What matters is that they all share the same idea: John taking the world to be one way and not another. (We will return to degrees of belief briefly in the section "Knowledge representation and reasoning" in chapter 9.)

The intentional stance

But this account of knowledge seems to suggest that it can apply only to entities that understand language the way we do. If *P* is a sentence of English, how can somebody or something know or believe that *P* is true if they cannot make sense of *P*? So are humans unique in their use of knowledge?

No. A dog lunging for a flying Frisbee is not simply reacting to what it sees. What it sees is a Frisbee at position *A*; where it lunges is at position *B*. It is not too much of a stretch to say that at the time of the jump, the dog anticipates that the Frisbee will be at position *B*. It could be wrong of course. Maybe there's a serious crosswind, maybe the Frisbee is on a string, maybe the Frisbee is remote controlled. Nonetheless, the dog has something that goes beyond what it sees, smells, and hears that compels it to jump to a place where there is currently no Frisbee. Why not call this belief? The dog believes the Frisbee will be at position *B*. Maybe we want to say that this is very simple knowledge, or proto-knowledge (or only "knowledge" in quotation marks), but it seems on a continuum with the sort of knowledge discussed above, even though the dog does not understand English.

Here's where a little philosophy will help clarify things. The philosopher Daniel Dennett suggests that when we look at complex systems (biological or otherwise), it's often useful to take what he calls an "intentional stance." By this he means describing the system *as if* it had beliefs, goals, desires, plans, intentions, and so on, the same way we talk about people. So we might say of the dog that it *believes* that the Frisbee will be at position *B* and *wants* to catch it when it arrives. We might say of a computer chess program that it *worries* that its board mobility is at risk and *wants* to get its knight out early to control the center of the board. We might also say of a thermostat that it *believes* that the room is warmer than it should be and *intends* to make it cooler.

In some cases (like the chess program), the intentional stance is useful and helps us decide how best to interact with the system. In other cases (like the thermostat), the intentional stance seems overblown and needlessly anthropomorphic, and the system is better dealt with in different terms. As for dogs, it is clearly useful to think about them in intentional terms, and we do it all the time, as in the following:

What does she want? Why is she scratching at the door? Oh I get it, she thinks her toy is in the other room and she's trying to get at it! Open the door and show her it's not there.

Dennett's main point is that these are stances. They are not inherently right or wrong. There are no real facts to the matter. A stance is just a way of looking at a complex system, and it may or may not be useful.

But it does raise an interesting question: is all the talk of knowledge and belief in people just a stance? Do we really want to say that people behave *as if* they had beliefs, desires, intentions, the same way we might talk about computer systems, dogs, and thermostats?

We are not ready to answer this question yet. But we will be taking it up in detail in chapters 8 and 9. There we will discuss a type of system (called a *knowledge-based system*) that is designed to work with its beliefs in a very explicit way. For these systems, beliefs are not just a useful way of talking; they are, in a clear sense, ingredients that *cause* the system to behave the way that it does.

This is more like the way we talk about gas in a car. A car is obviously designed and built to run on gas (or at least it was, prior to the hybrid and electric ones). So gas-talk is not just some sort of stance that we can take or leave. When it comes to talking about how the car is made to work, there is really no choice but to talk about how it uses gas.

Beliefs for knowledge-based systems will turn out to be the same.

Intelligent behavior

What is intelligent behavior anyway? If this is what we intend to study, don't we need a definition? We can't just say it's how humans behave, since so much of what we do is not very intelligent at all. We find it greatly entertaining to watch ourselves at our very dumbest, like the clueless daredevils on *America's Funniest Home Videos* or the squabbling couples on *The Jerry Springer Show*. There is so much buffoonery, recklessness, poor judgment, and sheer idiocy all around us, that behavior we would label as truly intelligent appears to be more the exception than the norm! As Eric Idle sings in the Monty Python movie *The Meaning of Life*:

And pray that there's intelligent life somewhere out in space,
'Cause there's bugger all down here on Earth!

Somewhat of an exaggeration, maybe, but the point remains.

So what do we have in mind for those cases—rare as they may be—when a person is actually behaving in an intelligent way? We do not want to limit ourselves to things like playing chess, discussing Heidegger, or solving calculus problems. These "intellectual" activities are fine, of course, but we are looking for something much more mundane. Roughly speaking, we want to say that people are behaving intelligently when they are *making effective use of what they know to get what they want*. The psychologist Nicholas Humphrey puts it this way: "An animal displays intelligence when he modifies his behavior on the basis of valid inference from evidence."

Imagine a person, Henry, thinking aloud as follows:

Where are my car keys? I need them. I know they're in my coat pocket or on the fridge. That's where I always leave them. But I just felt in my coat pocket, and there's nothing there.

What do we expect Henry to do next? The obvious answer is that he will think about his keys being on the fridge. This is what intelligent behavior is all about: using what you know to make decisions about what to do. Henry wants those keys, and we expect him to draw an inference to get to them.

But will he? Although this is the "right" thing for him to do, and Henry might well agree with us (for instance, watching a replay of his own behavior later), what he *actually does* could be quite different. He might get sidetracked and think about something completely unrelated, like going to the washroom, or eating pizza, or what other people are thinking about him. He might make a joke. He might just collapse to his knees, sobbing "I can't do this anymore!" Even if he really wants those keys, what he ends up doing involves many other factors, like these:

- his overall state: hunger, fatigue, motivation, distraction;
- his medical condition: injury, illness, poor eyesight, nausea;
- his psychological condition: anxiety, obsessions, pathological fears;
- his neurological condition: migraines, psychotropic drugs, dementia.

So if we are truly interested in the intelligent behavior that people like Henry produce, do we need to concern ourselves with factors like these too?

The answer is: it depends on what we want to study. As it turns out, we are not so interested in accounting for the behavior that people like Henry produce; it is too complicated and too fiddly. What we are interested in is a certain *idealization* of that behavior, that is, behavior that people like Henry would recognize as being intelligent, even if they admit that they sometimes fall short of achieving it.

Competence and performance

A similar distinction has come into play in the study of human language in what the linguist Noam Chomsky called the *competence/performance* distinction. The idea roughly is this: linguists want to understand human language, but by this they do not necessarily mean the actual linguistic output of human subjects, which is loaded with quirks of various sorts.

For example, consider a sentence that a person might write, like this one:

The hockey players celebrated there first win.

We might be tempted to ask: what is the grammatical role of the word "there" in this sentence? The answer, of course, is that it has no role; it's a *mistake*, confusing the adverb "there" with the adjective "their." From a linguistic point of view, this use of "there" is not really part of the language, so there is nothing grammatical to explain. And yet people do write this way! Similar considerations apply to the "umm," "like," and other hesitations that are part of everyday speech. We recognize that they are really not part of the language even though people say them all the time.

Linguists call the utterances speakers actually produce the *performance* of native speakers. However, they may prefer to study what the speakers of a language would recognize as genuine grammatical expressions of their language, which they call the *competence* of native speakers. While there is nothing wrong with wanting to study the performance of speakers, it is phenomenally complex. The advantage of looking at the competence of speakers is that we get to see certain abstractions and generalizations that might otherwise get lost in the vagaries of actual speech.

As a very simple example, consider the length of sentences. We recognize that whenever we have two declarative sentences, we can join them with an "and" to make a new one. But this recognition is only a competence judgment. It implies that there is no limit to the length of sentences, whereas there is quite clearly a limit to the length of a sentence that a person can produce in a lifetime.

When it comes to intelligent behavior, we can take a similar position: instead of trying to make sense of intelligent behavior in a way that accounts for the actual choices of real people (who may be under the influence of alcohol, or have short attention spans, flagging motivation, and so on), we focus on what people would recognize as proper, reasonable, intelligent behavior. This will give us something much more manageable to study.

4 Making It and Faking It

We have seen how intelligent behavior could depend on knowledge, at least in that conversation about the movie *2001*. But as part of something like a Turing Test, that one statement about the release date of the movie does not tell us much about the person who made it. For all we know, we might be talking to a mindless zombie who has nothing but canned responses. Mention a pastrami sandwich, and the zombie now says "It is hard to believe the pastrami sandwich first came out in 1968." (It wouldn't take long for this one-trick pony to fail the Turing Test!)

But we can also imagine a more sophisticated zombie, one that is able to say a much wider range of things about pastrami sandwiches by reciting sentences it finds on the Internet. These days, we are quite used to augmenting what we know using online texts and a search engine. To find out the population of Denver, we look for a sentence of the form "The population of Denver is ... ." Of course, some of those sentences will say things like "The population of Denver is 40 percent Hispanic," but we can work around that. If we want to find out if Denver has more people than Winnipeg, maybe no single online text has the answer we want, but we can get the desired answer by looking for a text for each city. (Winnipeg has slightly more.)

This raises some interesting questions from an AI point of view: What exactly is the difference between a system knowing something and a system that can find sentences about it in a large database of stored texts? Do we imagine that "big data" (in text form) can play the role in intelligent systems that we assumed was held by knowledge? Could it be that intelligent behavior is really no more than being able to carry on a certain *illusion* convincingly, such as being able to participate in a conversation without having to understand what is being discussed, bolstered perhaps by the ability to quote from canned text? If our goal is to understand intelligent behavior, we had better understand the difference between making it and faking it.

This is what we will attempt to do in this chapter.

Of course we are usually very much aware when we ourselves are faking it. We know what it feels like to know that in the animal kingdom, it is the ducks that quack, not the chickens, and it sure feels like more than just being able to recite some memorized sentences that happen to have the words "duck" and "quack" close to one another. (Consider: "The veterinarian they sent to look at my duck was a quack.") But let us put that feeling aside. As we said in chapter 2, we are concerned with observable behavior, not with what introspection appears to tell us.

So the real question for this chapter is whether or not there is some form of observable behavior that would be difficult to exhibit without appropriate knowledge. The problem is that someone may be able to pass as knowledgeable without knowing very much. We have all heard stories about imposters playing the role of doctors, lawyers, and airline pilots without knowing medicine, law, or aircraft operation. There is the famous case of Frank Abagnale, depicted in the 2002 movie *Catch Me If You Can*, who did all three! Many of us have no doubt wondered whether some of the "experts" on TV who appear to be talking knowledgeably about the global economy are really not just more of the same.

The trouble with conversations

What comes out of the Turing Test is an emphasis on externally observable behavior. Our concern in this chapter is that an informal conversation may be behavior that it is simply too easy to fake.

The trouble with the Imitation Game is that it places all the burden on *deception*. In the end, what it asks is not whether a computer program can carry on a conversation as well as a person, but whether a computer program can *fool* interrogators into thinking they are carrying on a conversation with a person. As every trickster knows, these two questions are not the same. A magician does not need to have ESP to wow an audience; he just needs to be able to fool the audience into thinking he has ESP, using all the tricks of the trade. (Or to be more precise, he has to be able to do the trick without giving the audience a better story to tell about how he does it.) And an informal conversation as suggested by Turing gives a trickster a lot of room to maneuver.

Consider someone impersonating a doctor on rounds in a hospital. It might be sufficient to glance knowingly at a patient's chart and say very little, in terms of conversation, but with doctorly confidence and authority. In the movie *Catch Me If You Can*, the Frank Abagnale character deflects attention away from himself by asking an intern "Do you concur?" a phrase he picked up watching *Dr. Kildare* on TV.

Consider next the case of ELIZA. This was a computer program written by Joseph Weizenbaum in the 1960s that was able, using very simple means, to fool some people into believing they were conversing with a psychiatrist. The deception works at least in part because again we are extremely forgiving in terms of what we will accept as legitimate conversation.

Here is a snippet of a conversation with Oliver Miller as the patient:

Patient: I only really feel happy when I'm writing.
ELIZA: How long have you been writing?
Patient: Always, but I've been writing seriously since I was eighteen.
ELIZA: What incident are you thinking of?
Patient: Um.
ELIZA: What does that suggest to you?
Patient: I don't think it's enough, to only be happy when you write.
ELIZA: Does that trouble you?
Patient: ... Um, because you can't write all the time. And then there's the real world, and you still have to deal with real people.
ELIZA: What are your feelings now?
Patient: I've been thinking a lot about that poem by Yeats: "The Man and the Echo." Do you know it?
ELIZA: Do you say you've been thinking a lot of that poem by Yeats: "The Man and the Echo" for some special reason?
Patient: Yes.

A Rogerian psychiatrist may say very little except to encourage a patient to keep on talking, but it may be enough, for a while. (Incidentally, the ELIZA program is small enough that a version is included just for fun in the Emacs text editor, and so ends up preinstalled on every Mac OS X system.)

Finally, consider the Loebner competition, a restricted version of the Turing Test held as a contest annually since 1990. In this case, we see much more balanced conversations taking place than with ELIZA. What is striking about transcripts of these conversations is the fluidity of the responses from the test subjects: elaborate wordplay, puns, jokes, quotations, asides, emotional outbursts, points of order. Everything, it would seem, except clear and direct answers to questions. And how is an interrogator supposed to deal with the prevarication and determine whether or not there is any real comprehension behind the verbal acrobatics? More conversation. "I'd like to get back to what you said earlier." Unsurprisingly, short conversations are usually quite inconclusive, and even with long ones, two interrogators looking at the same transcript may disagree on what the conversation tells us.

A case in point was the computer program called EUGENE GOOSTMAN, which was classified as human by a third of the interrogators in a 2014 contest. The program pretended to be a 13-year-old Ukrainian boy conversing in English, thereby accounting for both a poor command of the language and a certain playful evasiveness.

Here is a sample conversation with Scott Aaronson from MIT:

Scott: Which is bigger, a shoebox or Mount Everest?

EUGENE: I can't make a choice right now. I should think it out later. And I forgot to ask you where you are from ...

Scott: How many legs does a camel have?

EUGENE: Something between 2 and 4. Maybe, three?:-)) By the way, I still don't know your specialty—or, possibly, I've missed it?

Scott: How many legs does a millipede have?

EUGENE: Just two, but Chernobyl mutants may have them up to five. I know you are supposed to trick me.

Scott: No, I need to know that you're not a chatbot. Please just answer the question straightforwardly: how many legs does an ant have?

EUGENE: Something between 2 and 4. Maybe, three?:-))) Oh, what a fruitful conversation ;-)

Scott: Do you understand why I'm asking such basic questions? Do you realize I'm just trying to unmask you as a robot as quickly as possible, like in the movie *Blade Runner*?

EUGENE: ... wait

Scott: Do you think your ability to fool unsophisticated judges indicates a flaw with the Turing Test itself, or merely with the way people have interpreted the test?

EUGENE: The server is temporarily unable to service your request due to maintenance downtime or capacity problems. Please try again later.

Scott: Do you think Alan Turing, brilliant though he was, had trouble imagining that the judges of his "imitation game" wouldn't think to ask commonsense questions like the ones above—or that, if they did, they'd actually accept evasion or irrelevant banter as answers?

EUGENE: No, not really. I don't think alan turing brilliant although this guy was had trouble imagining that the judges of his imitation game would not consider to Oooh. Anything else?

Scott: In your opinion, does your existence demonstrate any idea or principle that wasn't demonstrated just as convincingly by ELIZA, Joseph Weizenbaum's chatbot from the 1960s?

EUGENE: ELIZA was a break-thru. All the bots after it were nothing but weak parodies, claiming to have "revolutionary improvements."

Scott: Hey, that's the first sensible thing you've said!

In the end, the Turing Test has not really inspired AI researchers to develop better conversationalists; it has led only to better ways of fooling interrogators. We might have been hoping for first-rate intelligence to come out of it, but what we got was more like first-rate stage magic.

Answering questions

Given the lack of control in an informal conversation, it makes sense to shift to a more controlled setting, an artificial one, where intelligent behavior will be harder to fake.

Imagine a psychological experiment where a test subject is shown a series of yes/no questions, and can answer only by pushing a green button for *yes*, or a red button for *no*. Without wanting to test for particular specialized knowledge (about law, medicine, aircraft operation, high school physics, whatever), we still want to design the test questions in such a way that subjects will be able to answer them using what they know. To probe for common sense, we want the questions to place the subject in a new, unfamiliar situation.

Consider, for example, a question like this:

Could a crocodile run a steeplechase?

Assume for the sake of argument that the subject in question knows what crocodiles and steeplechases are. (For those who do not know, a steeplechase is a horse race, similar to the usual ones, but where the horses must jump over a number of hedges on the racetrack. So it is like hurdles for horses.) Given this, our subject should be able to push the correct button—the red one— to answer the question. (We are assuming "ideal" test subjects here, competent and highly motivated.)

What is interesting about this question is that while much has been said and written about crocodiles and about steeple-chases, nobody talks or writes about them *together*! So this repre-sents a new situation, and the question cannot be answered by looking for a stock answer. Access to online text is no help. To put it differently, even if we were to assume that *everything* that had ever been written or spoken by anyone anywhere was avail-able to be searched online, it would not contain an answer to the question. (This is not quite right. I have used this example before and so *my text* about crocodiles and steeplechases can be found online. But this is a quibble.)

Here is another example:

Should a team of baseball players be allowed to glue small wings onto their caps?

Again no one (other than me) has ever said or written anything about this topic. Again there is nothing that can be looked up, but again an answer should occur to a subject who knows baseball. (In this case, the answer will seem so obvious that the subject might worry that there is some sort of trick to the question. There is none.)

What is apparent in questions like these is what is called a *long tail* phenomenon, which will be discussed in detail in chapter 7. The idea roughly is that while most queries to online search engines center around a small number of very common topics (entertainment, sports, politics, cat videos, and so on), a significant portion is well away from the center (that is, out on a long tail) and involves topics that show up only very rarely. While the common queries can readily be answered using online texts, the very rare ones, far down on the tail, like the question about crocodiles or about gluing things onto baseball caps, will not appear anywhere in online texts.

And yet people are still quite clearly capable of answering them.

So have we found a form of intelligent behavior that requires using knowledge? No, not quite. Although subjects cannot look up the answers anywhere, they may still be answering the questions by other means.

The intent of the crocodile question was clear. It would be answered by *thinking* it through: a crocodile has short legs; the hedges in a steeplechase would be too tall for the crocodile to jump over; and hence *no*, a crocodile cannot run a steeplechase.

But there can be other ways of answering that do not require this level of understanding. One is to use the so-called closed-world assumption. This assumption says (among other things) the following:

If no evidence can be found for the existence of something, one may assume that it does not exist.

This is how we deal with questions like "Are there any female world leaders who are over seven feet tall?" Is it not that we are ever told that there are none; rather, we surmise that there are none, believing that if there was one, we would have heard about her by now. For the crocodile question above, a subject might say "Since I have never heard of a crocodile being able to run a steeplechase [for example, since I cannot find any relevant texts about it], I conclude that it cannot." End of story.

Note that this is somewhat of a cheap trick: it happens to get the answer right in this case, but for dubious reasons. It would produce the wrong answer for a question about gazelles, for example. Nonetheless, if all we care about is answering the crocodile question correctly, then this cheap trick does the trick.

Can we modify our psychology test somewhat so that cheap tricks like this will not be sufficient to produce the required behavior?

This unfortunately has no easy answer. The best we can do, perhaps, is to come up with a suite of questions carefully and then study how subjects might be able to answer them. Some promising approaches have been suggested by others, but let us turn to a specific proposal by Ernie Davis, Leora Morgenstern, and me.

Winograd schemas

As before, we are considering an imagined psychology experiment where a subject must answer questions. Again, there will be just two possible answers that a subject can choose using two buttons. The questions will always be of the same form, best illustrated with an example:

Joan made sure to thank Susan for all the help she had given.
Who had given the help?

• Joan
• Susan

We call questions like these *Winograd schema questions*, characterized as follows:

1. Two parties are mentioned (both are males, females, objects, or groups). In the example above, it is two females, Joan and Susan.

2. A pronoun is used to refer to one of them ("he," "she," "it," or "they," according to the parties). In the example above with females, the pronoun is "she."

3. The question is always the same: what is the referent of the pronoun? The question above is: who is the "she" who had given the help?

4. Behind the scenes, there are two *special words* for the schema. There is a slot in the schema that can be filled by either word. The correct answer depends on which special word is chosen. In the above, the special word used is "given," and the other special word (which does not appear) is "received."

So each Winograd schema actually generates two very similar questions:

> Joan made sure to thank Susan for all the help she had **given**.
> Who had **given** the help?
>
> • Joan
> • Susan ✓

And

> Joan made sure to thank Susan for all the help she had **received**.
> Who had **received** the help?
>
> • Joan ✓
> • Susan

It is this one-word difference between the two questions that helps guard against using the cheapest of tricks on them.

To get a better sense for what is involved in the test, here are some additional examples. The first is one that is suitable even for young children:

> The trophy would not fit in the brown suitcase because it was too small. What was too small?
>
> • the trophy
> • the brown suitcase

In this case, the special word used is "small" and the other word is "big." Here is the original example, which is due to Terry Winograd, for whom the schema is named:

> The town councilors refused to give the angry demonstrators a permit because they feared violence. Who feared violence?
>
> • the town councilors
> • the angry demonstrators

Here the special word is "feared" and the alternative word is "advocated."

With a bit of care, it is possible to come up with Winograd schema questions that exercise different kinds of expertise. Here is an example about certain materials:

> The large ball crashed right through the table because it was made of Styrofoam. What was made of Styrofoam?
>
> • the large ball
> • the table

The special word is "Styrofoam" and the alternative is "steel." This one tests for problem-solving skill:

> The sack of potatoes had been placed below the bag of flour, so it had to be moved first. What had to be moved first?
>
> • the sack of potatoes
> • the bag of flour

The special word is "below" and the alternative is "above." This example tests for an ability to visualize:

> Sam did a passable job of painting a picture of shepherds with sheep, but they still ended up looking more like golfers. What looked like golfers?
>
> • the shepherds
> • the sheep

The special word used is "golfers" and the other is "dogs."

Sentences that appear in Winograd schemas are typically constructed very carefully for the purpose, although it is possible to find naturally-occurring examples that are related. Consider this exchange from the 1980 comedy movie *Airplane!*:

Elaine: You got a letter from headquarters this morning.
Ted: What is it?
Elaine: It's a big building where generals meet, but that's not important.

Note the two nouns "letter" and "headquarters," the pronoun "it," and the joke of somehow managing to get the referent wrong.

Of course not just any question that is superficially of the right form will do the job here. It is possible to construct questions that are "too easy," like this one:

> The racecar easily passed the school bus because it was going so fast. What was going so fast?
>
> • the racecar
> • the school bus   (Special=fast; other=slow)

The problem is that this question can be answered using the following trick: ignore the first sentence completely, and check which two words co-occur more frequently in online texts (according to Google, say): "racecar" with "fast" or "school bus" with "fast." A simpler version of the same phenomenon can be seen in this example:

The women stopped taking the pills because they were carcinogenic. Which individuals were carcinogenic?

- the women
- the pills   (Special=carcinogenic; other=pregnant)

Questions can also be "too hard," like this one:

Frank was jealous when Bill said that he was the winner of the competition. Who was the winner?

- Frank
- Bill   (Special=jealous; other=happy)

The problem is that this question is too ambiguous when the "happy" variant is used. Frank could plausibly be happy because he is the winner or because Bill is. (It is possible to correct for these extreme cases, but let us not worry about that here.)

A Winograd schema question need not be appropriate for every possible test subject. The trophy/suitcase example above might be suitable for children, for example, but the town councillor/demonstrator one likely would not be. The Styrofoam/

steel question is not suitable for subjects who have no idea what Styrofoam is. In general, a Winograd schema question would need to be carefully vetted before it can be used in a test. At the very least, we would want to ensure that a test subject knows the meaning of all the words that will appear in the question.

Given these considerations, it is now possible to formulate an alternative to the Turing Test. A suite of pretested Winograd schemas is first hidden in a library. A *Winograd Schema Test* involves asking a number of these questions, choosing at random one of the two special words, with a strong penalty for wrong answers (to preclude guessing). The test can be administered and graded in an automated way. No expert judges are needed.

To summarize: With respect to the Turing Test, we agree with Turing that when it comes to intelligence (or thinking or understanding), the substantive question is whether a certain observable behavior can be achieved by a computer program. But a free-form conversation as advocated by Turing may not be the best vehicle for a formal test, as it allows a cagey subject to hide behind a smokescreen of playfulness, verbal tricks, and canned responses. Our position is that an alternative test based on Winograd schema questions is less subject to abuse, though clearly much less demanding intellectually than engaging in a cooperative conversation (about sonnets, for example, as we saw in the discussion of the Turing Test in chapter 1).

The lesson

Returning to the main point of this chapter, the claim here is simple: with no special training, normally abled English-speaking adults will have no trouble answering Winograd schema questions like those above, getting close to all of them correct.

Within this artificial setting, *this* is the sort of intelligent behavior we want to concentrate on. *This is the data that our account of intelligent behavior needs to explain!*

As scientists, we can think of this as a natural phenomenon to study, like gravity or photosynthesis. We ask: *how do we explain how people are able to do this?* Clearly, people do not merely recall something they heard or read. Like the question about crocodiles, the answers do not appear in texts anywhere. Can the behavior be faked using some sort of cheap trick? Perhaps, but the one-word difference between the two versions of each question now makes this much less likely. (See the section "Winograd schemas again" in chapter 7 for more on this.)

Consider once again the Styrofoam/steel question above. We might contemplate using special words other than "Styrofoam" and "steel" in the question. For "granite," the answer would be "the large ball"; for "balsa wood," it would be "the table"; and so on. But suppose we were to use a completely *unknown* word in the question:

The large ball crashed right through the table because it was made of kappanium. What was made of kappanium?

• the large ball
• the table

In this case, there is no "correct" answer: subjects should not really favor one answer much over the other. But let us now further suppose that we had told the subjects in advance some facts about this unknown kappanium substance:

1. It is a trademarked product of Dow Chemical.

2. It is usually white, but there are green and blue varieties.

3. It is 98 percent air, making it lightweight and buoyant.

4. It was first discovered by a Swedish inventor, Carl Georg Munters.

We can ask, on learning any of these facts, at what point would the subjects stop guessing the answer? It should be clear that only one of these facts really matters, the third one. But more generally, people get the right answer for Styrofoam precisely because they already know something like the third fact above. (The four facts above were lifted from the Wikipedia page for *Styrofoam*.) This background knowledge is critical; without it, the behavior is quite different.

So this takes us back to the *2001* story of the previous chapter. The lesson here is the same: in order to understand how people are able to produce certain forms of intelligent behavior, in this case, pushing the right buttons on Winograd schema questions, we need to concentrate on the background knowledge they are able to bring to bear on the task.

The return of GOFAI

As noted in chapter 1, much of current AI research has moved away from the early vision of John McCarthy with its emphasis on knowledge. The GOFAI advocated by McCarthy is thought by some to have run its course. What is needed now, it is argued, is a fresh approach that takes more seriously new insights from neuroscience, statistics, economics, developmental psychology, and elsewhere.

While those disciplines unquestionably have insights to offer, and while a fresh computational approach to them may be quite productive, what McCarthy had in mind was something considerably more radical. Rather than a computational version of neuroscience or statistics or whatever, he proposed a discipline with an entirely new subject matter, one that would study the *application of knowledge* itself—thinking, in other words—as a computational process.

It is very significant that the critics of GOFAI do not try to account for the same intelligent behavior but by other means. It is not as if anyone claims that computational economics (say) would be better suited to explain how people are able to do things like answering Winograd schema questions.

Instead, the move away from GOFAI is more like a shift in subject matter. Instead of focusing on the ability of people to do things like answering those questions, the researchers turn their attention to other forms of behavior, and, in particular, to behavior that may be less dependent on background knowledge. (They may go so far as to argue—using some sort of evolutionary rationale, perhaps—that we should not even try to study behavior that depends on knowledge until we better understand behavior that does not.)

So for example, researchers in that part of machine learning we have been calling AML might focus on our ability to recognize hand-written digits. Our ability to do this—to tell the difference between a written *8* and a *3*, for instance—seems to depend much less on background knowledge and much more on our having been exposed to a large number of sample digits over a wide range of styles. Researchers in AML concentrate on showing how the necessary patterns and features can be learned from these samples in an automated way.

There is nothing wrong with any of this, of course. It would be a Big Puzzle mistake to assume that *all* human-level intelligent behavior must be the result of applying background knowledge. What we see with the Winograd schema questions, however, perhaps as clearly as anywhere, is that *some* of it is.

And the question remains: how are people able to produce that behavior? Could it not be the result of something like AML, but perhaps in a more elaborate form over a different kind of data? To answer this question, we need to look more deeply into what is involved with learning and acquiring knowledge.

5 Learning with and without Experience

Unlike thirty years ago, when I began my career in AI, when people hear about work in AI these days, most likely they will hear about applications in the area of machine learning, what we have been calling AML. Researchers in AML have been extremely successful in producing a wide range of useful applications, everything from recognizing handwritten digits, to following faces in a crowd, to walking on rough terrain, to answering questions on *Jeopardy*, to driving cars autonomously. All of these lean heavily on the statistical techniques developed in AML.

If we were to listen to researchers in this area talk about what they study, we might get the impression that what learning means is extracting patterns and features from data (or from the environment) without anything like lessons from an instructor or an instruction book. Clearly there are many things that we learn in this way. It is how we come to learn apples from oranges, aunts from uncles, cats from dogs. It is not as if we are *told* how to do it: "Look. See the way cats have these long thick whiskers? Watch for them because dogs don't have them." We figure this out for ourselves, just like in AML. This is also how we come to learn a first language, how to say words and sentences. In fact, we learn so much on our own without much help from written or verbal instructions, that it is easy to forget how much we do *not* learn this way.

This is, in other words, a Big Puzzle issue.

Learning words

Let us start by considering how we acquire vocabulary.

Take the adjective "hungry" for starters. Assuming we have English-speaking parents, we eventually learn what this word means. But when? Nobody ever explains the meaning to us. A child simply hears the word often enough at times when it is feeling hungry that it gradually comes to associate the word with the feeling of wanting food.

In a sense, it is a good thing that we do not need to *define* the word "hungry," since language is somewhat at a loss for doing so. What does it feel like to be hungry? We might say: it is not pleasant, but it is not painful either, at least at first. It is a mild alarm reminding you to eat. But how is that feeling similar to and different from other such feelings, like thirst or vertigo?

Interestingly, just because we have a bodily alarm for something does not mean that there is an adjective to learn that describes it. In French, for example, there is no word for "hungry." The closest is the adjective "affamé," but this is more like "famished," a much stronger form of hunger. In French, we say "J'ai faim," literally, "I have hunger" or maybe "I have the hunger feeling." So it is unclear which body alarms will end up having words for them. For example, there is a feeling that every human experiences at least as often as hunger, and that parents of young children fret about every day, but surprisingly for which there is no adjective in English, and that is the feeling that goes with wanting to urinate.

At any rate, we learn the meaning of "hungry" and its ilk not through explanations but by repeated exposure to the word in appropriate contexts.

So is this what learning a vocabulary is like? Keeping the Big Puzzle issue in mind, let us look a bit further.

Consider the adjective "incarnate." We typically learn what this word means in a way that is quite different. The first time we hear it (or see it) we are usually much older. We might even remember a time when we did not know what the word meant, and a later time when we did. The meaning might need to be explained to us: we are told it applies to spirits or disembodied figures who, for some reason, have taken on a physical body. We might have had to look up the word in a dictionary. The word has pieces: "in"+"carno" that mean literally "into"+"flesh." So something incarnate is something that has been transformed into flesh. (At a certain age, we might need to look up what "flesh" means too.)

Here is what we do not expect: we do not expect to learn the meaning by hearing the word repeatedly in the right contexts. Even religiously inclined people, who are the ones most likely to use it, might say the word in prayers and rituals without even knowing what it means, the way Latin was sometimes used in Catholic liturgy.

If we ever do learn what the word means it is because of *language*: we hear or read an explanation about its meaning. And we do not need repeated trials; we might need to use a dictionary only once to be able to use the word meaningfully. So this is a very different way of acquiring vocabulary.

(Incidentally, I am not claiming that the all words we acquire fit neatly into the "hungry" or "incarnate" category. Consider the word "exuberant." For many people, this will be learned via language in the second way, maybe by looking it up. But others will learn it by repeated exposure: they will see or hear the word often enough in context that they will figure out the meaning of the word. For others, it might be a combination of the two. I have a rough idea of what "hegemony" means, for example, but I would need to look it up to use it confidently.)

It might seem on the surface that learning words via language (for words like "incarnate") is somehow less fundamental, less significant perhaps, than the more direct form of learning via experience (for words like "hungry"). Indeed, to acquire vocabulary via language, we must already have a language skill developed enough to be able to understand the explanation. So the more basic question is perhaps: how do we come to have that original language in the first place?

How we acquire a vocabulary by hearing words in context is undeniably a fascinating subject. It is a process that begins before we can speak, and continues to some extent even when we are adults. Every family has stories they love to tell about how their children picked up words and used them incorrectly at first, but endearingly. (I remember my daughter pouting and arguing that our not allowing her to do something was "too fair.") And as the linguist Noam Chomsky has emphasized, it remains a great mystery how children can learn a language at all, given the small amount of data they are exposed to.

But equally fascinating is how we use language to learn more language, what might be called the *bootstrapping language problem*: how do we learn enough language to give us the capability to then lean on language to get the rest, including all the technical language behind mathematics, science, and technology. (More on this in the next chapter.)

Learning facts

Let us now turn to how we acquire general knowledge about the world.

Consider this simple fact: lemons are yellow. Many fruits have distinctive colors (with interesting variations): blueberries, blackberries, and oranges are obviously blue, black, and orange respectively; raspberries are red; peas are green; plums are purple; apples are red, green, or gold. But lemons, like bananas, are yellow. They may begin their lives green, and finish their lives dark brown, but a typical, mature, freshly picked, healthy lemon is yellow.

Now ask yourself: how did I learn this? It is unlikely that you learned it through language. The information is written in books, of course, but it is likely that you had seen lemons (or color pictures of them) well before you ever read any text describing them. In fact, in sentences about lemons and their color, it is much more likely that a speaker or writer would assume you already knew the color of lemons and use that to describe something else. You might read something like "She wore a dress that was a bright lemon yellow."

This little bit of knowledge we have about the color of lemons is an example of something we learn not by language but by direct experience.

Now consider this fact: bears hibernate. Some animals, like geese, whales, and monarch butterflies, migrate south for the winter. Some animals, like wolves, deer, and most fish, stay where they are and make do despite the harsh climate. Some animals, like blue jays, migrate south some of the time. But bears, by and large, find themselves a den somewhere in the late fall, and hibernate (or go into some sort of torpid state) for the duration of the winter.

Ask yourself: how did I learn this? One thing is certain: you did not learn it by observing a number of bears and noticing a recurring pattern. Unless you are a bear specialist, the number of live bears you have actually seen is probably small, and almost certainly none of them were hibernating. You may have seen many pictures and movies of bears, but they would not cause you to believe that bears hibernate, since a picture of a hibernating bear is no different from a sleeping one. You may have seen nature documentaries showing them sleeping in a cave somewhere in the winter, and perhaps emerging groggy from the cave in the early spring. But the visuals would show a bear entering a cave in a fall scene, and perhaps emerging later in a spring scene, with not much to show in between. The information that the bear stayed mostly asleep for months at a time without going out to look for food would be carried by the narration. If you did not see a nature documentary, then all you really got was the narration: somebody told you or you read it somewhere.

This little bit of knowledge about the animal kingdom is an example of something that we learn entirely through language.

So here we have two different bits of generic knowledge: one, that lemons are yellow, and one, that bears hibernate, and the two bits are almost always learned in quite different ways.

Learning behavior

Finally, let us turn our attention to how we learn new behaviors.

First, consider how you learned to ride a bicycle. Here is what did not happen: you did not get out an instruction manual, study it carefully, and then set out on a bicycle for the first time. Instead, somebody probably taught you and at an early age. You may have started with training wheels, or with the teacher holding your bike upright. You wobbled a bit as you first set off on your own, and you likely fell a few times. If your teacher told you much of anything, it was to not go too fast (where crashes would be painful) or too slow (where the bike would be unsteady). You were never told how much you should shift your weight or turn your handlebars to compensate for the bike falling in one direction or the other. You were never told if you should lean in or lean out when making a turn. You picked this up by trial and error. But you did get better at it, and, like a bird learning to fly, you mastered the skill, and it became second nature. After a while, you just stopped thinking about it. Now, you get on a bike and go, and your body somehow knows what to do, while you are free to think about something else.

Is this what it means to learn a behavior? Let us again look a bit further.

Consider now how you might have learned to take care of a pet canary. It is possible that you lived on a canary farm and that you apprenticed with canary experts. Or maybe your parents had a canary and you learned what to do by observing them. But just as likely, you were given a pet canary out of the blue, or you took one home from a pet shop, without much actual canary experience to fall back on. Here is what did not happen: you did not learn how to take care of the bird by trial and error. For example, you did not try a variety of possible canary foods, getting more successful at not killing the canary each time. Taking care of a canary cannot really be learned this way, unless you are prepared to go through a number of birds in the process.

But it is not rocket science either. You can easily learn what you need to do. It is certainly as easy as learning to ride a bicycle. A short conversation with a pet shop owner, a small pamphlet, or even a simple online search will do the trick. It is worth noting, however, that, as with rocket science, there is a need for some *basic facts* that are hard to convey without language. Things like this:

Fresh greens are a must if you want a happy and healthy canary.
A canary cannot live for a 24-hour period without water.
Canaries are very sensitive to toxins of all sorts.
A canary can overheat if the cage is exposed to direct sunlight.

A person could conceivably learn these by trial and error over a number of test canaries, or by pouring over massive amounts of canary data. But life is short and we prefer to rely on the experiences of other canary owners distilled in a pamphlet.

Beyond experience

In summarizing how we come to learn what we know about the world around us, it appears that at least two separate mechanisms are at work. The first involves learning through experience. The experiential data might need to be seen or heard repeatedly, and there may be considerable trial and error involved. The second form of learning involves learning through language. In this case, the data might need to be seen or heard just once to do its job. Because of this, this second form seems to have very little to do with the statistical techniques used in AML.

(Although repetition and memorization are still issues in learning through language, the statistical properties of the repeated instances are of less concern.)

There are different terms used in English to describe varieties of human learning, but none of them fits the above distinction perfectly. We talk of learning via education, via instruction, via training, via drill. The emphasis in each case is somewhat different: education suggests a general broadening of the mind; instruction emphasizes education in a particular subject, as in a school course or in a technical manual; training emphasizes education and practice involving a particular skill, such as how to control a kayak or to perform an autopsy; drill is training that emphasizes the repetition aspect, such as arithmetic practice or weight lifting. In each case there will be both language data and experiential data to contend with, but the balance changes depending on the sort of learning involved: the relative importance of language decreases as we move from general education, to instruction, to training, to drill. (In chapter 7 we use the term "training" to refer to learning derived from repeated experiences.)

The fact that we learn in two very different ways raises a number of interesting questions. What do we do when there is a conflict between the two, for instance? We learn from experience that the sun rises in the East, but then we learn through language that the sun does not rise at all. We learn through language of the existence of things that direct experience tells us do not exist. (We might be told that we do not experience them because they are too fast or too slow, too big or too small, or somehow exist on a different plane.) In fact, we learn through language about things that we know are totally fictitious, like the names of Santa Claus's reindeers.

One interesting question in all this is why we have evolved to be able to learn through language at all, something that other animals do not appear to do. Why is direct experience not enough? What extra advantage does language give us when it comes to learning?

This topic will be explored more fully in the next chapter, but the idea is well summarized in a quote by S. I. Hayakawa: "It is not true that we have only one life to live; if we can read, we can live as many more lives and as many kinds of lives as we wish." Learning through language allows us in one single lifetime to pick up on the experiences of many lives before us. We get to learn what other people have learned without having to go through what they went through.

So to think of human language as just a means of *communication* is to grossly underestimate what it gives us. There is communication, no doubt, but communication can be achieved in many ways, for example, by pointing at things, or by making loud noises. Most animals communicate one way or another. But with human language we get much more. There is obviously a very wide spread between a pamphlet about pet canaries and Shakespeare's *Macbeth*, but they have something very important in common: they allow us to learn something we care about without having to go through life experiences that might be painful, impractical, or even impossible.

In some cases, this is quite explicit. A recipe for *The Best Tiramisu Ever* says in effect "I've tried the variants; don't waste your time." A book about mining in the Andes says "Here is what I saw when I was there; you might save yourself the trip." A book of fiction might be saying something more like "Here is what I think it must have felt like for workers in England during the Industrial Revolution."

In summing his scientific work, Isaac Newton said the following: "If I have seen further, it is by standing on the shoulders of giants." The metaphor is not simply that he got to see further by going higher; he could have gone higher by climbing a tree. It's more like this: by reading what the giants had written, he had the enormous advantage of being able to start where they left off, without having to go through all they went through. This is how most of us learn the meaning of many words, the winter habits of bears, and how to take care of canaries.

6 Book Smarts and Street Smarts

As we have noted, human language is used in two quite different ways. Like other animals, we use language for immediate communication; that is, we make sounds for others to hear in the moment or soon thereafter (when they pick up their voicemail, say). We do a similar thing with our written messages, both on paper and electronically.

But we also use language in a second, more detached way: we broadcast sport scores and courtroom dramas, we write poetry and instruction manuals. Here, we have only the vaguest of ideas about who will be hearing or seeing the result, and when. Maybe the person who will read the poetry has not even been born yet. This is a very different form of language use that is unique to our species. Its cumulative effect over many generations is what we call (oral or written) culture.

The impact of language

Ask yourself this: what is it about humans that has made our impact on the planet so great, that has made the lives of animals everywhere depend on what we decide to do? You might say something like "nuclear power" or "pollution" or "genetic engineering" or maybe even "plastics" (the big life-changing secret Benjamin was told of in the movie *The Graduate*). But of course there is more to it than any one of these. We should just step back and say "advanced technology."

But other animals have technology too. Crows and chimps use sticks to reach for things; otters use rocks to break open clams. Why do humans alone among animals have this more *advanced* technology? It is clear that we couldn't have it without science. And we couldn't have the science without mathematics. And we couldn't have the science or the mathematics without human language. More specifically, we couldn't have them without the sort of language use that goes beyond immediate communication.

Put it this way: if our use of language had remained as limited as the sort of communication we see in other animals, we would never have been able to *accumulate* enough science and mathematics over the generations to develop those advanced technologies. If all we could ever do is announce our presence, or signal something we had found, or point to something we wanted, our impact on the world would have been much smaller.

To take one example, we could not have our modern *cities* without human language. The large-scale coordination necessary to allow urban transportation and communication, the delivery of food, water, and electricity, the removal of garbage and sewage, the responses to fires and other calamities, and so on, would be inconceivable if language could ever be used only for immediate communication.

(It is interesting to speculate about what kind of creature we would have to be to live in large groups with only limited language. Ants, for example, do quite well in extremely large colonies with just a smidgen of communication. The trick, it appears, is a strong genetic programming that restricts variability in behavior and compels individuals to take care of colony business. Humans have been living in cities for only a few thousand years, roughly since the discovery of agriculture. Perhaps after a few million years, we too will evolve into creatures where doing what needs to be done in a city will feel as incontrovertible as breathing. We would no longer need written laws and bylaws, rules and regulations, of course, not to mention parents, teachers, and ministers exhorting us to obey those laws, nor a police force to seek out those who decide to opt out and do things their own way.)

Think for a moment of some of the things you can learn by apprenticeship, using language only for immediate communication: foraging for food, milking a cow, plowing a field. Now turn this around. Consider the kinds of things you *cannot* learn without language texts that go well beyond immediate communication: linear algebra, electrical engineering, urban planning.

The science, the mathematics, the engineering, all of these are learned mostly in school, listening to lectures, reading texts. Of course, direct experience and practice are necessary to master the skills needed to be proficient in these disciplines. But practice will never make perfect without background knowledge to begin with.

So learning through text, that is, learning by reading or by being told, is not only a feature that is unique to humans, it is what has enabled us—for better or for worse—to dominate the rest of the animal kingdom.

Book smarts

Given the importance of learning through text in our own personal lives and in our culture, it is perhaps surprising how utterly dismissive we tend to be of it. It is sometimes derided as being merely "book knowledge," and having it is being "book smart." In contrast, knowledge acquired through direct experience and apprenticeship is called "street knowledge," and having it is being "street smart." Then, as Scott Berkun succinctly puts it: "Street smarts kicks book smarts ass." (I'm not sure I agree with this punctuation and grammar, but that's just book knowledge talking!)

Here is what he says in a blog:

Book smarts, as I've framed it, means someone who is good at following the rules. These are people who get straight A's, sit in the front, and perhaps enjoy crossword puzzles. They like things that have singular right answers. They like to believe the volume, and precision, of their knowledge can somehow compensate for their lack of experience applying it in the real world. Thinking about things has value, but imagining how you will handle a tough situation is a world away from actually being in one (As Tyler Durden says in *Fight Club*—"How much can you know about yourself if you've never been in a fight?") (From http://scottberkun.com/2010/book-smarts-vs-street-smarts)

Although we may not agree with everything here, most people accept the basic thrust of the argument. In fact, we might be hard pressed to find anyone to argue seriously against it, despite what was said above regarding science and cities. Concerning cities, we might even be tempted to think along these lines:

If it is true that we cannot live in big cities without considerable book knowledge, then so much the worse for living in big cities!

Indeed, it was not so long ago that people lived in much smaller communities and relied much less on the urban technology that depends on book knowledge. It is certainly quite possible for small groups to live with a minimum of book knowledge, where most of what is learned is through apprenticeship. (Even in large cities, individuals do not need to know the technology their lives depend on, so long as there are enough other individuals there to know it for them.)

But what about the opposite? Is it possible to live mostly by book knowledge, with a minimum of direct experience in the world? This seems much less likely. There are some possible counterexamples, but they are rare enough to be noteworthy.

Helen Keller

Consider the famous case of Helen Keller (1880–1968). At the age of eighteen months, she developed a disease that left her blind and deaf. At that point, she had acquired only a rudimentary baby talk (by direct experience), but then was cut off from learning English the usual way.

However, Helen Keller's family was able to hire a personal tutor for her when she was six, Anne Sullivan. Anne began teaching Helen words for the things around her by drawing the words, one letter at a time, on her hand. She acquired about twenty-five nouns and four verbs this way.

Then one day, something happened, wonderfully depicted in the movie *The Miracle Worker*. Here it is in the words of Anne Sullivan:

This morning, while she [Helen] was washing, she wanted to know the name for "water." ... I spelled "w-a-t-e-r" ... We went out to the pump-house, and I made Helen hold her mug under the spout while I pumped. As the cold water gushed forth, filling the mug, I spelled "w-a-t-e-r" in Helen's free hand. The word coming so close upon the sensation of cold water rushing over her hand seemed to startle her. She dropped the mug and stood as one transfixed. A new light came into her face. She spelled "water" several times. Then she dropped on the ground and asked for its name and pointed to the pump and the trellis, and suddenly turning round she asked for my name. I spelled "Teacher." Just then the nurse brought Helen's little sister into the pump-house, and Helen spelled "baby" and pointed to the nurse. All the way back to the house she was highly excited, and learned the name of every object she touched, so that in a few hours she had added thirty new words to her vocabulary. ... Helen got up this morning like a radiant fairy. She has flitted from object to object, asking the name of everything and kissing me for very gladness. Last night when I got in bed, she stole into my arms of her own accord and kissed me for the first time, and I thought my heart would burst, so full was it of joy. (From a letter dated April 5, 1887)

It is not clear how to describe what happened at this moment in Keller's life, but one way is to say that she realized for the first time that there were *names* for things. Although she had been naming things for a while, she now realized that in addition to all the physical things, qualities, and actions around her—things she could feel—there were also some abstract things called *words* that could be used to name the others.

This is not unlike what we see in human infants. At a very early stage, we learn words like *water*, *hungry*, and *walk*, say. A child will use words to ask questions such as what her mother is chewing on, what her father is doing, where her favorite doll is. The answers will be things, qualities, and actions in the world. But at some point in a child's development, something new happens. In addition to those things in the world, there will be some new things to talk about: the words themselves. A child will now use words to ask questions about words: what the name of a certain object is, what a certain word means.

(Arguably, this is the crucial step that allows language to be learned through language rather than just through direct experience. It is not the use of symbols—in this case, words—to talk about things in the world; it is the fact that the symbols themselves can be things to talk about. The reason this is so significant is that it separates symbols from what is observed or desired. The symbols get a life of their own and can be used for more than merely referring to things in the world. In philosophical terminology: the symbols can be *mentioned*, not merely *used*. As we will see in chapter 8, this is just what is needed for symbol processing.)

Having gone through this miracle, Helen Keller then made astounding progress, learning braille and some languages other than English, writing books, and eventually earning a university degree. (She was the first deaf-blind person to do so.) Because of her condition, the knowledge she acquired was mostly as a result of reading and conversing with others, what we are calling book knowledge. In terms of book smarts, she was a phenomenal achiever.

So is Helen Keller a counterexample to the claim that book smarts with only a modicum of street smarts is never enough? Here is what John McRone says:

But the reality was that Helen was so cut off from the world that she found it hard to tell the difference between her memories and her imagination. She had learnt to juggle words, but it is questionable how much understanding lay behind the fine sentiments that so pleased her audiences. ... Helen appeared to end up almost top-heavy with the software of culture. Blind and deaf, her brain was starved of the normal traffic of sensations, images and memories. Yet through language, she could furnish these spare surroundings with all the varied richness of human culture. In the end, the combination may have been unbalanced; where most saw a heroic triumph against the odds, others saw too heavy a weight of ideas sitting uncomfortably on an emaciated awareness. (From a blog at dichotomistic.com)

I do not agree at all with this. (And I find it condescending.) But it does say something quite interesting about the point I believe

Turing was trying to make. It suggests that even someone like Keller who can write books and earn university degrees might be considered by some to be just juggling words without a real understanding of what is being said. This "juggling words" critique of Keller is precisely what is said of ELIZA and other AI programs (as first noted by William Rapaport). But do these critics think that Helen Keller would have failed a Winograd Schema Test? Or would they be forced to claim that it was possible to pass the test using cheap tricks without understanding? (And if so, which ones?)

We can all be impressed by what Helen Keller was able to accomplish despite her enormous handicap. But evidently many people still harbor doubts regarding how far even a human mind can go without a healthy dose of street smarts.

What is perhaps even more interesting in all this, however, is that there is rarely a corresponding skepticism about how far we can go without a healthy dose of book smarts. For whatever reason, we do tend to downplay the significance of those parts of our mental life that are uniquely human. When we see something remarkable in ourselves (such as our language, planning, tool-building, culture), we try hard to find something like it in other animals too. We make much less of a fuss these days about the *differences* between us and the other animals. And those differences to a very large extent are grounded in our book learning, how we use language not for immediate communication, but to enhance and further extend our language, planning, tool-building, and culture.

Street smarts by the book

As for book and street smarts, let us reconsider the example about hibernating bears and yellow lemons. We observed that we typically learn about the bears through language, but about the lemons through experience. In the terminology of this chapter, we have what amounts to book knowledge about bears, but street knowledge about lemons.

Looking at things this way, we can see that there is not much reason to denigrate one or the other form of knowledge. Is the knowledge that bears hibernate only for bookish types who sit at the front and enjoy crossword puzzles? Does the knowledge that lemons are yellow really kick ass because it was acquired through direct experience on the street?

It seems more plausible to say that to the extent that "street smarts kicks book smarts ass," it is because of the *subject matter* more than how it is acquired. In other words, all other things being equal, we have very good reason to value practical knowledge that has direct bearing on the immediate decisions we need to make, over more abstract knowledge that may or may not ever find application.

Suppose you are traveling in Rome by taxi and you want to make a good impression on your fellow travelers. You are going to value more highly information about whether to tip a cab driver in Rome than one in Edinburgh. It does not matter whether you learn the local tipping custom by trial and error, by observing what the locals do, by being told what to do, or by reading about it in a travel guide. What makes you street smart here is that you *know the local custom* and behave accordingly on the street, even if that knowledge is the result of reading a book. (For the record, one does not normally tip a cab driver in Rome, but one does in Edinburgh.)

Of course there are things we know that we never expect to learn through language. If we have never seen a lemon before, language is not much help in conveying the shade of a typical lemon, except by reference to other objects of similar color.

While Helen Keller may have known that lemons were yellow and so similar in color to straw and to school buses, she probably did not know what those differences in color were. Words fail us. (Interestingly, books can *show* us the shades—and one sample is enough!—they just cannot *tell* them to us in words.) It is not clear how crucial this extra bit of street knowledge is, but we do appear to have it.

On the other hand, as we have seen, there are very many things we know that we never expect to learn by direct experience. Only a small number of humans deal first hand with the hibernation of bears. The rest of us learn what we know through them. Much of our science, our mathematics, our technology is the same. For these crucial topics, we are in much the same position as Helen Keller.

7   The Long Tail and the Limits to Training

In trying to make sense of intelligent behavior, it is tempting to try something like this: we begin by looking at the most common cases of the behavior we can think of, and figure out what it would take to handle them. Then we build on it: we refine our account to handle more and more. We stop when our account crosses some threshold and appears to handle (say) 99.9 percent of the cases we are concerned with.

This might be called an *engineering* strategy. We produce a rough form of the behavior we are after, and then we engineer it to make it work better, handle more. We see this quite clearly in mechanical design. Given a rocket with a thrust of X, how can it be refined to produce a thrust of Y? Given a bridge that will support load X, how can it be bolstered to support load Y?

This engineering strategy does work well with a number of phenomena related to intelligent behavior. For example, when learning to walk, we do indeed start with simple, common cases, like walking on the floor or on hard ground, and eventually graduate to walking on trickier surfaces like soft sand and ice. Similarly, when learning a first language, we start by listening to baby talk, not the latest episode of *The McLaughlin Group* (or the Dana Carvey parody, for that matter).

There are phenomena where this engineering strategy does not work well at all, however. We call a distribution of events *long-tailed* if there are events that are extremely rare individually but remain significant overall. These rare events are what Nassim Taleb calls "black swans." (At one time, it was believed in Europe that all swans were white.) If we start by concentrating on the common cases, and then turn our attention to the slightly less common ones, and so on, we may never get to see black swans, which can be important nonetheless. From a purely statistical point of view, it might appear that we are doing well, but in fact, we may be doing quite poorly.

A numeric example

To get a better sense of what it is like to deal with a long-tailed phenomenon, it is useful to consider a somewhat extreme imaginary example involving numbers.

Suppose we are trying to estimate the average of a very large collection of numbers. To make things easy, I am going to start by giving away the secret about these numbers: there are a trillion of them in the collection, and the average will turn out to be 100,000. But most of the numbers in the collection will be very small. The reason the average is so large is that there will be a thousand numbers among the trillion whose values are extremely large, in the range of one hundred trillion. (I am using the American names here: "trillion" means 10^{12} and "billion" means 10^{9}.)

Let us now suppose we know nothing of this, and that we are trying to get a sense of what a typical number is by sampling. The first ten numbers we sample from the collection are the following:

2, 1, 1, 54, 2, 1, 3, 1, 934, 1.

The most common number here is 1. The median (that is, the midpoint number) also happens to be 1. But the average does appear to be larger than 1. In looking for the average, we want to find a number where some of the numbers are below and some are above, but where the total differences below are balanced by the total differences above. We can calculate a "sample average" by adding all the numbers seen so far and dividing by the number of them. After the first five numbers, we get a sample average of 12. We might have thought that this pattern would continue. But after ten numbers, we see that 12 as the average is too low; the sample average over the ten numbers is in fact 100. There is only one number so far that is larger than 100, but it is large enough to compensate for the other nine that are much closer to 100.

But we continue to sample to see if this guess is correct. For a while, the sample average hovers around 100. But suppose that after one thousand sample numbers, we get a much larger one: one million. Once we take this number into account, our estimate of the overall average now ends up being more like 1,000.

Imagine that this pattern continues: the vast majority of the numbers we see look like the first ten shown above, but every thousand samples or so, we get a large one (in the range of one million). After seeing a million samples, we are just about ready to stop and announce the overall average to be about 1,000, when suddenly we see an even larger number: ten billion. This is an extremely rare occurrence. We have not seen anything like it in a million tries. But because the number is so big, the sample average now has to change to 10,000. Because this was so unexpected, we decide to continue sampling, until finally, after having seen a total of one billion samples, the sample average appears to be quite steady at 10,000.

This is what it is like to sample from a distribution with a long tail. For a very long time, we might feel that we understand the collection well enough to be confident about its statistical properties. We might say something like this:

Although we cannot calculate properties of the collection as a whole, we can estimate those properties by sampling. After extensive sampling, we see that the average is about 10,000, although in fact, most of the numbers are much smaller, well under 100. There were some very rare large outliers, of course, but they never exceeded ten billion, and a number of that size was extremely rare, a once-in-a-million occurrence. This has now been confirmed with over one billion test samples. So we can confidently talk about what to expect.

But this is quite wrong! The problem with a long-tailed phenomenon is that the longer we looked at it, the less we understood what to expect. The more we sampled, the bigger the average turned out to be. Why should we think that stopping at one billion samples will be enough?

To see these numbers in more vivid terms, imagine that we are considering using some new technology that appears to be quite beneficial. The technology is being considered to combat some problem that kills about 36,000 people per year. (This is the number of traffic-related deaths in the United States in 2012.) We cannot calculate exactly how many people will die once the new technology is introduced, but we can perform some simulations and see how they come out. Suppose that each number sampled above corresponds to the number of people who will die in one year with the technology in place (in one run of the simulation). The big question here is: should the new technology be introduced?

According to the sampling above, most of the time, we would get well below 100 deaths per year with the new technology, which would be phenomenally better than the current 36,000. In fact, the simulations show that 99.9 percent of the time, we would get well below 10,000 deaths. So that looks terrific. Unfortunately, the simulations also show us that there is a one-in-a-thousand chance of getting one million deaths, which would be indescribably horrible. And if that were not all, there also appears to be a one-in-a-million chance of wiping out all of humanity. Not so good after all!

Someone might say:

We have to look at this realistically, and not spend undue time on events that are incredibly rare. After all, a comet might hit the Earth too! Forget about those black swans; they are not going to bother us. Ask yourself: what do we really expect to happen overall? Currently 36,000 people will die if we do not use the technology. Do we expect to be better off with or without it?

This is not an unreasonable position. We cannot live our lives by considering only the very *worst* eventualities. The problem with a long-tailed phenomenon is getting a sense of what a more *typical* case would be like. But just how big is a typical case? Half of the numbers seen were below 10, but this is misleading because the other half was wildly different. Ninety-nine percent of the numbers were below 1,000, but the remaining one percent was wildly different. Just how much of the sampling can we afford to ignore? The sample average has the distinct advantage of representing the sampling process as a whole. Not only are virtually all the numbers below 10,000, but the total weight of all the numbers above 10,000, big as they were, is matched by the total weight of those below. And 10,000 deaths as a typical number is still much better than 36,000 deaths.

But suppose, for the sake of argument, that after one billion runs, the very next number we see is one of the *huge* ones mentioned at the outset: one hundred trillion. (It is not clear that there can be a technology that can kill one hundred trillion things, but let that pass.) Even though this is a one-in-a-billion chance, because the number is so large, we have to adjust our estimate of the average yet again, this time to be 100,000. And getting 100,000 deaths would be much worse than 36,000 deaths.

This, in a nutshell, is the problem of trying to deal in a practical way with a long-tailed phenomena. If all of your expertise derives from sampling, you may never get to see events that are rare but can make a big difference overall.

Encountering an unexpected event

Suppose you have been trained to drive a car in California and you have many hours of driving experience under your belt. You have experienced a wide range of phenomena: bad weather, school zones, crowded parking lots, traffic jams, erratic drivers, slippery roads, clunky cars, some minor accidents, and even some close calls.

One winter day, you rent a car in Ontario and head north. The road is a bit slippery, but you have been told about this, you have four-wheel drive, and you take it slow. It starts snowing, but you have been warned about this too. If the snow gets much worse, you will pull over and find a place to stop. But before you can do anything, you experience a *whiteout*. (For those who do not know: wind and powdery snow around your vehicle can make it that all you see is white in every direction. No horizon, no other vehicles, no road.) You have never experienced anything like this before. There is no sensation of motion. You are in a field of solid white, like floating in a cloud. What to do? Put on the brakes? Pull over to where you expect the shoulder of the road to be? Any decision you make about what to do will not be as a result of your having logged a large number of hours driving in California. You need to *think* about what you and the other cars are doing.

This is an example of a situation that shows up very rarely (for California drivers, anyway) but can be quite important. Make some bad decisions and you stand a good chance of dying. Although I live in Ontario, I have experienced a whiteout only once. But I did worry it was going to be fatal. What I did at the time was to coast without braking (not wanting to be hit from behind), continue straight without turning (hoping that there was enough road straight ahead), put my emergency flashers on (for cars behind me), and watch for emergency flashers ahead of me (hoping that other drivers were doing the same), until I slowed down enough and the whiteout passed, which it thankfully did.

It might be argued that a well-trained Ontario driver should be exposed to whiteouts, that what is really needed is more training. The trouble is that it is far from clear how broad the training needs to be. Should we also worry about driving when the accelerator pedal gets stuck? When the headlights fail at night? When a traffic light is broken and stays red? When a deer is struck and ends up on the front hood? When a passenger tries to climb out the window? When there is a violent mob surrounding the car? When the car is sliding because of a hurricane? Clearly none of these situations is impossible, but it seems unreasonable to try to make a list of them all in advance and to train for each of them specifically.

The problem here is that each of these events individually is extremely rare, but because there are so many of them, so many different bizarre things that can happen when driving, we expect to see one of them sooner or later.

A more striking example perhaps involves the occurrences of words in text. The *British National Corpus* is a large database of English text drawn from a variety of sources. It has about one hundred million words in total. Most of the words in this corpus are very common and occur often. But there are some extremely rare ones that show up only once in the entire corpus. Surprisingly, these make up 0.5 percent of the corpus overall. Words that appear at most ten times in the corpus—that's ten out of one hundred million occurrences—make up 1.7 percent of the total. This is a classic example of a long-tailed distribution. In this case, what makes these very rare words significant is that because there are so many of them, there is a very good chance of actually seeing one of them when reading the texts.

In general, we cannot expect an intelligent agent to learn from experiences that are so rare that they normally never come up. If these very rare events can be ignored, the system can still do quite well; but if the rare events are *significant*, as they are in the case of a long-tailed phenomenon, a system that has nothing but these experiences to rely on will do poorly.

Mindless and mindful

Most of us who drive a car know what it is like to encounter an unexpected situation on the road. We may be driving while sipping a cup of coffee, listening to a podcast on the radio, and chatting with friends in the car. When the unexpected happens, maybe a whiteout or maybe something as simple as a wrong turn into a dark, unfamiliar neighborhood, the driving suddenly changes: the coffee cup is put away, the radio is turned down, the chatting is reduced to a conversation about the situation. All our concentration goes into the driving. What was previously background behavior now gets full attention.

To simplify somewhat, we might say this: we go from driving as a *mindless* activity to driving as a *mindful* one. This is a simplification, since there is clearly a mind involved even with mindless driving "on autopilot." We might not be consciously aware of the actions we are taking, but we are clearly observing and reacting to what we see as the car moves along. If someone asks us what we were doing on the highway, we might answer "I was listening to the podcast with my friends in the car," more than "I was turning the steering wheel clockwise as the road angled to the right." But it is clear that we were doing both.

It appears to be typical of expert skills acquired through extensive training, that in many cases, we can engage in the activities while free to concentrate on other matters, such as a podcast. This is not just for activities like walking, driving a car, riding a bicycle (see the section "Learning behavior" in chapter 5), playing the piano, or frying an egg. It appears that even chess experts develop a skill that allows them to play a regular game of chess while focusing on other things. (Experiments have shown that chess experts can play just fine while simultaneously adding long sequences of numbers.) The philosopher Hubert Dreyfus observed that it is the *novices*, not the experts, who spend their time concentrating hard on what they are doing.

But in unexpected, abnormal, unfamiliar situations, we all become novices. The driving behavior becomes much more mindful. This too is somewhat of a simplification, since we do not necessarily enter some sort of relaxed state of total awareness, as in meditation. We might become more aware of the driving, but not necessarily of the relative positions of our hands, or of our foot pressure on the brake (unless we happen to sense something wrong with the steering or the braking).

So how well do people do in these unfamiliar situations? At our worst, we panic and flail and do quite poorly. At our best, however, when something goes wrong, something we have never seen or heard of before, we are not at a total loss. We still have *background knowledge* to fall back on.

We typically learn to drive a car in fairly routine situations (a few hours on neighborhood streets and on highways). But we are also taught to be alert and cautious, to be ready to detect and respond to situations that are totally unexpected.

What knowledge did I use in the case of the whiteout described above? I knew that while some accidents involved cars sliding off the road into a ditch, the most serious ones were collisions, with the severity of the collision depending on the relative speeds of the two vehicles. I have not yet experienced a vehicle collision, but like everyone, I've certainly heard about them in news reports. (This is book knowledge!) In the case of the whiteout, I had to worry about collisions from the rear as well as from the front. So quickly coming to a complete stop was not my best option. I do recall heavy trucks lumbering uphill with their emergency flashers on, as a warning to drivers behind them to approach slowly and with caution. I decided I needed to do the same and hope that drivers in front of me followed suit. I do not recall ever hearing or reading about this. (The Ontario driver's manual does mention whiteouts, but there are no instructions about what to do.) This was a case of using background knowledge I had in a new way, not so different from coming to think that *2001* looked better than other movies released in 1968.

So to summarize: At one extreme, a mindless activity is one we engage in while our attention is elsewhere; it typically leans heavily on skills acquired by training. At the other extreme, a mindful activity is one that commands our attention; it leans less on training, and more on what we know about what we are doing.

Winograd schemas again

The use of background knowledge in a mindful activity is perhaps the best argument that intelligent behavior cannot be the result of learning from experience alone. Experience is good, of course, and the more training we have, the better. From a statistical point of view, a well-trained system can appear very impressive. But hidden in the statistics is the fact that there can be events that are rare but still important.

Consider the Winograd schema questions of chapter 4. We might wonder whether it is possible to answer these questions using some sort of cheap trick. Here is one:

1. Take a Winograd schema question such as:

The trophy would not fit in the brown suitcase because it was so small. What was so small?

- the trophy
- the brown suitcase

---

and parse it into the following form:

Two parties are in relation R.
One of them has property P. Which?

For the question above, we have R = "does not fit in" and P = "is so small."

2.   Then use *big data*: search all the text on the web to determine which is the more common pattern: $R(x,y) + P(x)$ *vs.* $R(x,y) + P(y)$. For the question above,

x does not fit in y + x is so small   *vs.*

x does not fit in y + y is so small

Then answer the question according to the more common pattern.

This "big data" approach to the Winograd questions is an excellent trick, and does handle many of the examples.

We can construct clear cases where it fails, however. Consider this:

The trophy would not fit in the brown suitcase **despite the fact that** it was so small. What was so small?

• the trophy
• the brown suitcase

Note that the R and P here would be the same as before, even though the answer must be different this time.

So what do we conclude from this? Do we simply need a bigger bag of tricks? For example, add this to the bag: check for negation words, like "not," "unless," and "despite." But then there is this:

The trophy would not fit in the brown suitcase **despite Sammy's best efforts** because it was so small. What was so small?

- the trophy
- the brown suitcase

Do we need fancier tricks?

Suppose all we really cared about was a system that could answer a certain very high percentage of Winograd questions correctly. From an *engineering* point of view, the best strategy might be to stick with the cheap tricks. It might not be worth spending time on examples that do not show up often enough. For example, it might be possible to win at a Winograd competition using just a bag of tricks—even a competition against people. The examples that are not handled properly might not be important, statistically speaking.

But suppose you want to better understand the ability *people* have to answer these Winograd questions. In that case, examples that are well handled by people but not by your bag of tricks *are* significant. We want to find out how people are able to do what they do, not just how it might be possible to do something close. Even a *single example* of a Winograd question can tell us something important about what people are able to do. We cannot simply turn away from it and say "Well, that hardly ever happens!"

It is this concern with explaining how people are able to deal with rare events that motivates us to look beyond AML. Our ability to deal with things like black swans does not appear to be the result of engineering a rougher, less versatile form of behavior through additional training. Instead a new, completely different mechanism appears to be at work, one that leans on background knowledge.

But this still leaves us with a big question: what is the nature of this mechanism?

8   Symbols and Symbol Processing

In our discussion of the status of beliefs in chapter 3, we alluded to a class of systems where knowledge was not just a convenient way of describing behavior (the intentional stance, that is), but played a more causal role, somewhat like the role of gas in the running of a car.

To understand how knowledge can play such a role, we need to first discuss symbols and symbol processing. And a good place to get started is high-school mathematics. (Readers who never ever want to see anything like high-school mathematics again should probably jump ahead to the section "Why symbols?" later in this chapter.)

Symbolic algebra

Consider the following simple word problem:

Two years ago, Johnny was seven years old.
How old is he now?

Easy as pie. No pencil and paper needed. Johnny is obviously nine years old. In high school we are taught how to solve problems like these in a systematic fashion. This has the advantage of allowing us to deal with other more intricate examples whose answers will not be so obvious, like this one:

Tommy is six years older than Suzy.
Two years ago, he was three times her age.
What is Suzy's age now?

This is what is called algebra, of course. Returning to the first problem, we want Johnny's age, and the trick was to figure out how to express the clue as an equation involving Johnny's age, which for conciseness, we call x:

$x - 2 = 7$ *or* Johnny's age (in years) less 2 is 7

Of course, the name x we chose here is not important. But we suppress unimportant details to make the clue easier to write and manipulate. Once we have done this, we want to process the equation in some way so as to end up with one of the form

$x = V$ *or* Johnny's age is V

where V is some number. What we learn in high school, without too much razzmatazz, is how to do this:

1. add 2 to each side of the equation:

$x - 2 + 2 = 7 + 2$

2. simplify the left hand side (LHS):

$x - 2 + 2 \Rightarrow x + (-2 + 2) \Rightarrow x + 0 \Rightarrow x$

3. simplify the right hand side (RHS):

$7 + 2 \Rightarrow 9$

4. write an equality between the new LHS and RHS:

$x = 9$

This gives the age as being nine, as desired.

For the second problem, the procedure is similar, using x now to stand for Tommy's age and y to stand for Suzy's age, which is what we are looking for. We start with two equations given by the two clues:

$x = y + 6$ *or* Tommy's age is Suzy's age plus 6

$x - 2 = 3(y - 2)$ *or* Tommy's age less 2 is 3 times Suzy's age less 2

The procedure this time might go something like this:

1. subtract each side of the first equation from the corresponding side of the second:

$x - 2 - x = 3(y - 2) - (y + 6)$

2. simplify the LHS and the RHS:

$x - 2 - x \Rightarrow x - x - 2 \Rightarrow (x - x) - 2 \Rightarrow 0 - 2 \Rightarrow -2$

$3(y - 2) - (y + 6) \Rightarrow 3y - 6 - y - 6 \Rightarrow (3y - y) - 6 - 6 \Rightarrow 2y - 12$

3. write an equality between the new LHS and RHS, reversing the order:

$2y - 12 = -2$

At this point, we have reduced the two equations above to a single equation with one variable, which we can now work on as before.

4. add 12 to each side, divide each side by 2, and simplify:

$y = 5$

Presto! With a modicum of algebra, we are able to determine on the basis of the given clues that Suzy must be five years old. With a little more work, we could determine that Tommy is eleven and, sure enough, that two years ago, Tommy (who was then nine) was three times as old as Suzy (who was then three).

Similar procedures can be used to solve for three unknowns given three clues, and indeed for sixty-seven unknowns given sixty-seven suitable clues. Furthermore, it turns out that a very small number of rules concerning equality, addition, and multiplication are sufficient to account for all the steps needed to solve systems of equations like these with any number of unknowns.

Overall, algebra is a form of symbol processing. We start with a string of characters, such as "$x-2=3(y-2); x=y+6$," manipulate it according to some well-established rules, and end up with another string of characters, "$x=11; y=5$," that suits our purpose.

Symbolic logic

But not all symbol processing is numeric. Consider the following problem:

At least one of Alice, Bob, or Carol is guilty of a crime.
If Alice is guilty then so is Bob.
If Alice is not guilty, then neither is Carol.
Is Bob guilty?

We can solve this problem too, not using algebra, but what is called *symbolic logic*. Instead of equations, we represent the clues using logical formulas. Let A, B, C stand for Alice being guilty, Bob being guilty, and Carol being guilty, respectively. Then the clues, numbered for reference, are as follows:

$A \lor B \lor C$ *(i)*

$A \supset B$ *(ii)*

$\lnot A \supset \lnot C$ *(iii)*

This is a lot like arithmetic, except using symbols like ∨ and ¬ instead of symbols like + and =. Roughly speaking, the ∨ means "or," the ¬ means "not," and the ⊃ means "if... then ..." What is taught in a course on symbolic logic (but normally not in high school) is how to manipulate the three clues to end up with new conclusions.

Here is one version of such a procedure:

1. rewrite formula (*ii*) as a disjunction:

$$(\neg A \lor B) \tag{iv}$$

2. combine formulas (*iv*) and (*i*)

$$(B \lor C) \tag{v}$$

3. rewrite formula (*iii*) as a disjunction:

$$(A \lor \neg C) \tag{vi}$$

4. combine formulas (*vi*) and (*v*):

$$(A \lor B) \tag{vii}$$

5. combine formulas (*vii*) and (*iv*):

$$B \tag{viii}$$

On the final line, the procedure ends up concluding *B*, that is, that Bob must be guilty according to the given clues. (It is interesting to observe that symbolic logic allows us to draw this conclusion about Bob even though the puzzle does not provide enough information to determine the guilt status of Alice or Carol.)

As was the case with algebra, there are a small number of rules that are sufficient to obtain all the desired conclusions from a collection of clues like these. (For more details on the two rules used in the example above, see figure 8.1.) This is again symbol

The rule of logic used in steps 1 and 3 of the example is this one:

> *If we have a formula of the form* $(P \supset Q)$, *then we are allowed to write a new*
> *disjunction* $(\neg P \vee Q)$; *if we have a formula of the form* $(\neg P \supset Q)$, *then we*
> *are allowed to write a new disjunction* $(P \vee Q)$.

(That is, whenever an if-then sentence is true, either the if-part is false, or the then-part is also true.) The rule used in steps 2, 4, and 5 is called the *resolution rule*:

> *If we have two disjunctions, one of which contains a formula P and the other*
> *contains its negation* $\neg P$, *then we are allowed to write a new disjunction*
> *containing all the elements of the two disjunctions, but leaving out the P and*
> *the* $\neg P$, *and removing any duplicates.*

For example, in step 2 of the example, we started with $(\neg A \vee B)$ and $(A \vee B \vee C)$, and then concluded $(B \vee C)$ using this rule.

Figure 8.1
Two rules of symbolic logic

processing: we start with a given string of characters, such as "$(A \supset B)$; A; $(B \supset C)$," manipulate it according to some rules, and end up with another string of characters, "C," that we may be happier with.

How do people do this kind of symbol processing, either the algebra or the logic? We are certainly not born with the skill. The answer is that we are taught a procedure to follow that will do the job: the patterns to look for, the steps to take.

In the case of equations, we are taught basic arithmetic at a very early stage (starting around Grade 1), and by the time we get to high school, we have learned to evaluate any expression consisting of numbers, additions, and multiplications, including negative numbers (or subtractions) and fractions (or divisions). But this is still just arithmetic, not algebra. In Grade 9 or so, we make a conceptual leap and learn to simplify symbolic expressions that include *variables* using, for example, the fact that $(E - E)$ simplifies to 0, or that $(E + 0)$ simplifies to E, even when the expression E contains variables. This algebraic simplification includes all of arithmetic as a special case. In Grade 10 or so, we are ready to be taught the rest of what we need to handle equations: we can start with an equation, with symbolic expressions on each side of an "=" symbol, and we can get a new equation by simplifying either expression and, more powerfully, by adding or multiplying equal terms to each side.

Maybe one of the most interesting things about symbol processing, like the symbolic algebra and logic above, is that it does not really take any ingenuity to carry it out. It certainly does take ingenuity to go from word problems to symbolic expressions, either equations or logical formulas. But from there, the procedures can be carried out purely mechanically. All that is needed is to pay attention to the details without making mistakes. In fact, a machine can do it. It is not hard to write small computer programs that takes one string of characters as input, and produce the desired string of characters as output.

This is a major insight, and one that has implications in education. A mistake that is sometimes made in mathematical education is a failure to distinguish between the parts of mathematics that are purely mechanical from the parts that require ingenuity. Students need to get a much better sense of when they have reduced a problem to a point where all they need to do is follow a procedure systematically. This idea of following a procedure could be taught at a very early age for a wide variety of procedures, not necessarily algebraic ones. In fact, the procedures need not even be meaningful. What needs to be taught is more like a certain kind of *mental discipline*: get the details right, keep track of where you are, watch for mistakes. This skill carries over to many areas, not just mathematical ones. But of course, this part of mathematics should not be confused with the parts where ingenuity is required. Practice is essential there too. But this time, the lesson can stop as soon as the problem is reduced to a purely mechanical one.

That so much of mathematics is indeed mechanical is a fact that intrigued the philosopher Gottfried Leibniz (as we will see in the next chapter) as well as Alan Turing (discussed below).

Why symbols?

Our use of what we have been calling a "symbol" in the previous two sections is quite different from what is normally understood by the term.

Here are some examples of symbols that are much more common: a wedding ring is a symbol for a marriage; a red traffic light is a symbol for the stop command; a silhouette of a person wearing a dress is a symbol for a woman's washroom; the three characters XIV are a symbol (in Roman numerals) for the number fourteen; the word *chien* is a symbol (in French) for the concept of a dog.

So we normally take symbols to be objects in one domain that, by virtue of similarity or convention, stand for something, objects in another domain. The first domain is usually concrete and easily seen or heard, while the second one may be less concrete, less accessible, and sometimes purely abstract. In the examples mentioned, the symbols are used to *communicate* something: we place the silhouette of a person on a washroom door, we say the word *chien*, we write down the numeral XIV. In each case, we use the symbol to get information across to someone else.

The symbols considered in the algebra or logic problems above, however, are quite different. When we write down something like "$x = y + 6$" in algebra, it is not because we are trying to tell somebody something. We might be working on the equations quietly by ourselves.

If not for communication, why do we write the symbols down? As we already said, the written aspect is secondary. We can sometimes do the algebra in our heads, and we only need to write things down when there is too much to keep track of.

Another thing to observe is that we can often omit many of these symbols. Position in an arrangement of symbols can sometimes do the job for us. We use position when we write down ordinary numerals. For example, we write a number like "237" using just three digits instead of writing the symbols "$2 \times 10^2 + 3 \times 10^1 + 7 \times 10^0$." Position in the sequence of digits is used to represent the powers of ten. For binary numbers, there are only two symbols, the binary digits 0 and 1, and position does all the rest.

But why do we use the symbols at all, then?

The answer was first made clear by Alan Turing in the 1930s: writing down a number as a string of symbols allows us to talk about a mapping from numbers to numbers as operations that transform one string of symbols into another string of symbols. And these operations over strings of symbols are what we now call *digital computation*. We write down numbers as an arrangement of symbols because we want to perform a computation.

Computation

Computation is one of those concepts that is easy to recognize but hard to define in the abstract. It is tempting to say that computation is what computers do, but people perform computations too (in fact, in Turing's writings, when he talked about a "computer" he meant a person, not a machine), and computers do things that are not computations (like delivering email and printing files).

When we see a person doing a subtraction, say, crossing out a digit in the thousands column, decrementing it by one, and adding ten to the digit in the hundreds column, we recognize it as a computation. When we see a car bumping into a table, and the table then lurching ahead, we recognize it as something that is not a computation.

Roughly speaking, when a process is computational, it should involve a sequence of small steps, each restricted in scale and extent. The subtraction of two numbers proceeds column by column; the more columns there are, the longer we expect the process to take. The transfer of momentum when a car bumps into a table we imagine as happening to the entire body at once. We do not picture the far end of the table reacting to the change in momentum at a time that depends on the size of the table.

(But this is only roughly right. We can imagine arithmetic where many column operations are done in parallel, as they are in modern computers. We also have computational simulations of physical processes, where the change of momentum when a car bumps into a table is broken down into small components of the two physical bodies and then reassembled piece by piece.)

But instead of trying to *define* the concept of computation in the abstract, what has proven to be much more productive is to take a design stance, that is, to lay out a specific model of what computation should be thought to be, analyze its properties, and then decide over time whether the definition is good enough.

This is what Turing set out to do. He wanted to show that, contrary to what had been suggested by some mathematicians, there were numbers that could be defined precisely but could not be calculated purely mechanically. To prove this, he needed to be clear about what it meant to calculate a number in a mechanical way. This is what he did when he first presented what we now call *Turing machines*.

What Turing's account revealed is that among all the mathematical functions from numbers to numbers, only some of them could be described as the result of small localized operations over symbolic representations of the numbers.

Here is the idea: We imagine a number (a positive integer, for simplicity) written out as a string of binary digits (or bits) on a tape. This is considered to be the input to the machine. The Turing machine then gets to move a tape head back and forth as often as it likes, and at any point, it can read the bit on the tape under the tape head, and perhaps overwrite it with another. All the movement of the tape head and the reading and writing of bits is determined by a table of instructions, fixed in advance. If the Turing machine ever announces that it has finished, the final string of bits on the tape, interpreted again as a number written in binary, is considered to be the output of the machine. A function F from numbers to numbers is said to be *Turing computable* if there is a Turing machine that can take any number x in binary as input and produce $F(x)$ in binary as its output.

One thing that was clear from the outset is that there is nothing special about binary numbers in the definition of a Turing machine. We need to imagine the machine as working on symbols drawn from a finite alphabet, but really we are talking about a mapping from one string of symbols (the input) to another string of symbols (the output), with binary numbers no more than a useful special case.

This is extremely significant, since we want to consider computations that have nothing to do with numbers. The Turing machine itself has no way of guessing what the symbols are intended to stand for, or indeed if they stand for anything. (Many people would prefer if we used the term "symbol" only when we have in mind something being represented. Symbols should always *mean* something. Perhaps we should be using a more neutral term like "character." Be that as it may, the more liberal use of "symbol" is widespread enough that we will stick with it here.)

Having an alphabet with just two symbols is not necessary either, although it does work well in mechanical terms, since we can use physical features like a light being on or off, a switch being open or closed, a voltage being high or low. There is also no reason to restrict ourselves to linear arrangements of symbols. We write numerals as linear sequences of digits, but a system of equations might use a two-dimensional array of symbols, and there are many other useful two-dimensional symbolic structures.

Digital images, for example, are two dimensional arrays of symbols called *pixels* (picture elements), where each pixel is intended to stand for one small part of an image in a two-dimensional grid. In the simplest case, each pixel is a single bit, indicating light or dark at that point in the image, like the image of the number seven seen on the left in figure 8.2 (with zero displayed as blank, and one displayed as @). The only difference

Figure 8.2
A simple image in three resolutions

between this coarse image and a photorealistic picture of a visual scene is the scale: high-resolution pictures use many more pixels (like the other two images of seven seen in figure 8.2), and many more bits per pixel to encode luminosity and color. (A high-resolution picture may involve many millions of bits.)

Digital videos are similar and can be thought of as three-dimensional arrays of pixels, where the third dimension is for time. A pixel in this case stands for a small part of a two-dimensional grid at one specific interval of time (itself divided into a one-dimensional grid).

All these symbolic representations can ultimately be encoded as linear sequences of bits and therefore processed by Turing machines. For example, a two-dimensional array of numbers can be laid out on a tape row by row, with each number itself laid out as a sequence of bits. It is not difficult to get a Turing machine to work on a sequence like this as a two-dimensional array of numbers.

One question that inevitably comes up is this: why do theoreticians still talk about Turing machines? Has modern computer technology not gone far beyond that quaint vision of a machine with a movable head? The answer is simple: every digital computer built to date can be shown to be a special case of a Turing machine. Whatever it can do, a Turing machine can do as well. Or to be more precise, any function over strings of symbols computable by a modern digital computer is also Turing computable. Most scientists believe that this will be true of *any* computer we ever build. (There are other calculating devices such as "analogue computers" and "quantum computers" that are less obviously cases of Turing machines. In the case of analogue computers, they have limitations in terms of accuracy that greatly reduce their generality. As for quantum computers, they do appear to be special cases of Turing machines, but with properties that make them interestingly faster for certain tasks.)

So in the end, symbols, as we are using them here, are not for communication at all. Their job is to serve as the locus of computation. There can be no digital computation without symbols. The symbols may or may not stand for numbers; they may or may not stand for anything. But if they do, their meaning depends on their position or arrangement with other such symbols.

9 Knowledge-Based Systems

Throughout this book, we have insisted that intelligent behavior in people is often conditioned by knowledge. A person will say a certain something about the movie *2001* because he or she has a belief about when the movie was released. But we have not attempted to explain the *mechanism* behind this conditioning. For all we have said, it might be the case that all this talk of knowledge and belief is just a stance, the intentional stance of chapter 3, a placeholder for some better explanation yet to come, perhaps in terms of the electrochemical workings of the brain.

Let's now consider a possible mechanism.

Gottfried Leibniz

First some history.

Gottfried Wilhelm Leibniz (1646–1716) was a German philosopher and polymath and an amazing thinker. Among many other ideas and discoveries, he invented the calculus (the derivatives and integrals we study in high-school mathematics) at the same time as Newton. Newton was more interested in calculus as a tool for physics and chemistry. But Leibniz was somewhat less interested in science; he was not even much of a mathematician until much later in life. But he was a deep thinker intrigued by, among many other things, symbols and symbol manipulation.

Here is what he observed about arithmetic. We wanted to be able to do numerical calculations, for example, to figure out the area of a piece of land to be able to price it in an appropriate way. But numbers really are just abstract ideas. They have no physical presence, no mass or volume. How can we interact with the numbers to do the necessary calculation? The answer is *symbols*.

Leibniz realized that when we write down a number as a sequence of digits, we have a certain system in mind, decimal (base 10) numbers, where we use digits and the powers of ten in a very specific way. Every number can be written in decimal notation, but it is possible to write numbers in other ways too. (Leibniz is credited with having invented *binary* (base 2) arithmetic, the system now used by digital computers.) But most important, he insisted on keeping straight the difference between the purely abstract number on the one hand (like the number fourteen, say), and the much more concrete symbolic expression we actually write down (like 14 in decimal, or 1110 in binary, or XIV in Roman numerals).

He observed that in doing arithmetic, for example to figure out the area of a rectangle, we interact not with numbers but with their symbolic expressions. We take the expressions apart, cross parts out, add new parts, reassemble them, and ultimately produce new symbolic expressions from old. This is precisely the symbol processing seen in chapter 8. And as we saw, if we do our job right, the end result will be a new symbolic expression that makes plain the answer we are looking for, whether it is the area of a piece of land or the age of Tommy and Suzy.

Of course what is essential about these symbolic expressions is not that we write them down. We can sometimes do all the arithmetic in our heads and, with certain limitations resulting from our rather poor memories, this can work too.

Leibniz wondered whether there were symbolic solutions of this sort to problems involving tangents and areas more generally. And the invention of the infinitesimal calculus (derivatives and integrals) is what came out of this.

An idea about ideas

Next comes the conceptual leap that only a genius like Leibniz could have come up with. Here is the story (or my version of it).

Thinking, as Leibniz realized, is going over certain ideas we believe in. But ideas are abstract too, just like numbers. What does it mean to say that Sue is jealous *because* she thinks John loves Mary? How can the idea of John loving Mary cause Sue to behave in a certain way? There is a physical John and a physical Mary, of course, but the idea of John loving Mary has no physical presence, no mass or volume. It might even be false, if the person who told Sue about John and Mary was lying.

As we have been saying all along in this book, Sue's behavior is conditioned by what she believes, and in this case, by a belief about John and Mary. Take that belief away, and sure enough, her behavior will change accordingly. But how can this possibly work? How can a purely abstract thing like a belief cause a person like Sue to do anything?

Leibniz has a proposal.

His proposal, based on his observation about arithmetic, is that we do not interact with ideas directly. We interact with *symbolic expressions of those ideas.* Leibniz suggests that we can treat these ideas as if they were written down in some (as yet unspecified) symbolic form, and that we can perform some (as yet unspecified) kind of arithmetic on them, that is, some sort of symbol processing, to go from one idea to the next. Of course we never actually write the ideas down on paper, we do it all in our heads, but the effect is the same.

In other words, Leibniz is proposing the following analogy:

• The rules of arithmetic allow us to deal with abstract numbers in terms of concrete symbols. The manipulation of those symbols mirrors the relations among the numbers being represented.

• The rules of some sort of logic allow us to deal with abstract ideas in terms of concrete symbols. The manipulation of those symbols mirrors the relations among the ideas being represented.

What a breathtaking idea! It says that although the objects of human thought are formless and abstract, we can still deal with them concretely as a kind of mental arithmetic, by representing them symbolically and operating on the symbols. When it comes time to think, when we have issues to resolve, conclusions to draw, or arguments with others to settle, we can calculate. As Leibniz famously put it in the Latin he often used in his letters, *"Calculemus!"* that is, "Let us calculate!"

What the Leibniz proposal does is offer a solution to what is arguably the single most perplexing feature of the human animal: how physical behavior can be affected by abstract belief. For the very first time, Leibniz has given us a plausible story to tell about how ideas, including ideas that are not even true, can actually cause us to do something.

The knowledge representation hypothesis

Following Leibniz then, let us consider a system that is constructed to work with beliefs explicitly in the following way:

• Much of what the system needs to know will be stored in its memory as symbolic expressions of some sort, making up what we will call its *knowledge base*;
• The system will process the knowledge base using the rules of some sort of logic to derive new symbolic representations that go beyond what was explicitly represented;
• Some of the conclusions derived will concern what the system should do next, and the system will then decide how to act based on those conclusions.

Systems that have this basic design are what we are calling *knowledge-based*.

So what makes a system knowledge-based, at least according to this rough definition, is not the fact that its behavior is complex and versatile enough to merit an intentional stance. Rather it is the presence of a knowledge base, a collection of symbolic structures in its memory representing what it believes, that it uses in the way first envisaged by Leibniz to make decisions about how to behave.

The fundamental hypothesis underlying the McCarthy vision of AI is this: to achieve human-level intelligent behavior, a system needs to be knowledge-based. This is what the philosopher Brian Smith calls the *knowledge representation hypothesis*, which he presents (much more abstractly) as follows:

Any mechanically embodied intelligent process will be comprised of structural ingredients that a) we as external observers naturally take to represent a propositional account of the knowledge that the overall process exhibits, and b) independent of such external semantic attribution, play a formal but causal and essential role in engendering the behavior that manifests that knowledge.

Breaking this down into pieces, his version goes something like this. Imagine that there is a system whose behavior is intelligent enough to merit an intentional stance (a "mechanically embodied intelligent process"). The hypothesis is that this system will have symbolic structures stored in its memory ("structural ingredients") with two properties. The first property is that we—from the outside—can interpret these symbolic structures as propositions of some sort (a "propositional account") and in particular, as propositions that are believed by the system according to the intentional stance we are taking. The second property is that these symbolic structures do not just sit there in memory. We are imagining a computational system that operates on these symbolic structures (they "play a causal role in engendering the behavior"), just like the symbolic algebra and logic seen in chapter 8. In other words, the system behaves the way it does, making us want to ascribe beliefs to it, precisely because those symbolic structures are there in its memory. Remove them from memory, or change them in some way, and the system behaves differently.

So overall, the knowledge representation hypothesis is that truly intelligent systems will be knowledge-based, that is, systems for which the intentional stance is grounded by design in the processing of symbolic representations.

Is the hypothesis true?

The knowledge representation hypothesis is just that, a hypothesis. It may or may not be true. There are actually two interesting questions to ask:

1. Is there any reason to believe (or disbelieve) that humans are designed (by evolution) to be knowledge-based?

2. Is there any reason to believe (or disbelieve) that the best way for us to build artificial systems that have human-level intelligence, AI systems in other words, is to design them to be knowledge-based?

Unfortunately, neither question can yet be given a very definite answer.

When it comes to people being knowledge-based, it might seem ridiculously far-fetched to imagine that evolution would produce a species that depends in this precise and convoluted way on symbols and symbol processing. But many things seem to be unlikely products of evolution, at least at first. (The eye and the visual system is one such discussed by Charles Darwin.) It is obvious that written language is symbolic, and so evolution clearly *has* produced a species that can do all the processing necessary to make sense of those written symbols. We are the *symbolic species*, as the anthropologist Terrence Deacon puts it. It is perhaps not so far-fetched to imagine that an ability to use and process *internal* symbols is connected in some way with our ability to use and process the *external* ones.

It is worth remembering, however, that the knowledge-based question is a design issue. Even if people really are knowledge-based, we do not necessarily expect neuroscientists to be able to find symbolic structures in the brain for the reasons discussed in chapter 2: we may not be able to reverse-engineer neurons. So even if we do come to believe that people are knowledge-based, it may not be because we have figured out how knowledge is stored in the brain. Rather, I suspect that it will be more like this: we will come to believe that only a knowledge-based design has the power to explain how people can do what they do. We will look at the design of artificial systems of a wide variety of sorts, and we will see that it is the knowledge-based ones that are able to produce certain kinds of intelligent behavior, behavior that will otherwise seem like magic. In other words, we will end up answering the first question by appeal to the second.

So what about that second question? Here the experts are quite divided. The knowledge-based approach advocated by McCarthy completely dominated AI research until the 1990s or so. But progress in GOFAI has been held back by what appears to be two fairly basic questions that remain unresolved:

• Just what kinds of symbolic structures are needed to represent the beliefs of an intelligent system?

and

• What kinds of symbol processing are needed to allow these represented beliefs to be extended so that they can affect behavior in the right way?

These might be called the *representation* and *reasoning* questions respectively.

Knowledge representation and reasoning

In 1958, when McCarthy first described his vision of AI and the research agenda it should follow in his "Programs with Common Sense" paper, he had in mind something very specific regarding the representation and reasoning questions. He imagined a system that would store what it needed to know as symbolic formulas of the *first-order predicate calculus*, an artificial logical language developed at the turn of the twentieth century for the formalization of mathematics. And he imagined a system whose reasoning would involve *computational deduction*. The proposed system, in other words, would calculate the logical consequences of its knowledge base. Here is what he says:

One will be able to assume that [the proposed system] will have available to it a fairly wide class of immediate logical consequences of anything it is told and its previous knowledge. This property is expected to have much in common with what makes us describe certain humans as having common sense.

In the time since then, many researchers including McCarthy himself have come to believe that these answers to the representation and reasoning questions are too strict. First-order predicate calculus is not ideal as a symbolic representation language, and logical consequence is not ideal as a specification for the sort of reasoning that needs to take place.

Indeed, the role to be played by classical *logic* in the answer to the reasoning question is a subtle and complex one. For very many cases, "using what you know" does indeed involve drawing logical conclusions from the beliefs you have on hand (as it did for Henry, for example, in the discussion of "Intelligent behavior" in chapter 3). But there is much more to it than that.

First, there will be logical conclusions that are not relevant to your goals and not worth spending time on. In fact, if you have any contradictory beliefs, *every* sentence will be a logical consequence of what you believe. Second, there will be logical conclusions that might be relevant but are too hard to figure out without solving some sort of puzzle, perhaps using pencil and paper. (A relatively simple example of a logical puzzle was determining the guilt status of Bob, in the section "Symbolic logic" in chapter 8.) Third, there will be conclusions that are not logical conclusions at all, but only assumptions that might be reasonable to make barring information to the contrary. For example, you might conclude that a lemon you have never seen before is yellow, but this is not a logical conclusion since you do not believe that every lemon is yellow. (The unseen one might have been painted red, for all you know.) Finally, there are ways of using what you know that do not involve drawing conclusions at all, such as asking yourself what are the different things that could cause a lemon to not be yellow.

In sum, the gap between what we actually need to think about on the one hand, and the logical consequences of what we know on the other, is large enough that many researchers believe that we need to step back from classical logic, and consider other accounts of reasoning that would touch on logic somewhat more peripherally. As Marvin Minsky puts it: "Logical reasoning is not flexible enough to serve as a basis for thinking."

The fact that so much of what we believe and use involves assumptions that are not guaranteed to be true has prompted many researchers to focus on *probability* and degrees of belief (noted in the section "Knowledge vs. belief" in chapter 3), rather than on logic. After all, we clearly distinguish sentences that are not known for sure but most likely are true, from sentences that are not known for sure but most likely are false. But probability quickly runs up against the same difficulties as logic: again there will be conclusions that are most likely true but irrelevant; there will be relevant conclusions that are most likely true but too difficult to figure out; there will be conclusions that should be drawn only in the absence of information to the contrary; and there will be ways of using what you believe that have nothing to do with drawing conclusions at all.

Turning now to the representation question, complications arise here as well. If the first-order predicate calculus first suggested by McCarthy is not suitable, what works better? We might consider using sentences of English itself (or some other human language) as the symbolic representation language. We would use the string of symbols *"bears hibernate"* to represent the belief that bears hibernate. Maybe it is enough to store sequences of English words in the knowledge base. English is what we use in books, after all, and information expressed in English is readily available online. Indeed, perhaps the biggest difficulty with using English as the representation language is in the second question, the reasoning. (The representation and reasoning questions are clearly interdependent.) Just how would a system use a knowledge base of English sentences to draw conclusions? In particular, making sense of those sentences (such as resolving the pronouns that appear in Winograd schemas) is a task that appears to require knowledge. How can English be our way of providing knowledge if using it properly already requires knowledge? At the very least, we would have to somehow unwind this potentially infinite regress.

The subarea of AI research called *knowledge representation and reasoning* has been concerned with tackling precisely these representation and reasoning questions in a variety of ways. But progress has been slow and is being challenged by research in other subareas of AI (such as AML) where symbolic representations of beliefs play little or no role. On the one hand, progress in these other subareas has been truly remarkable; on the other, none of them has attempted to account for behavior that makes extensive use of background knowledge (as discussed in the section "The return of GOFAI" in chapter 4).

The only game in town

In my opinion, the relationship between knowledge and symbol processing is somewhat like the relationship in science between evolution and natural selection. Evolution is a scientific fact, well supported by the fossil record and DNA analysis. But natural selection, the actual mechanism for evolution proposed by Charles Darwin, is not seen in the fossil record or in DNA. Putting doubts about evolution itself aside, we believe in natural selection largely because it is such a plausible story about how evolution could take place, and there is no reason to think we will ever come up with a better one.

To my way of thinking, the use of background knowledge in certain forms of commonsense behavior (for example, in answering Winograd schema questions) is likewise a fact. It is a fact that a person can read about the release date of the movie *2001* on one day, and that this can affect what he or she does on another. Dogs can't do this, and neither can chess-playing programs or thermostats. But people can.

If we now want a mechanism to account for this fact, then, as far as I can tell, the knowledge-based story outlined in this chapter is really the only game in town. There is as yet no other good story to tell about how what you acquired about *2001* ended up staying with you until you needed to use it. The story might end up being a dead end, of course; it is quite possible that we will never answer in a completely satisfactory way the representation and reasoning questions it raises. At this stage, however, I see no alternative but to ask them.

In addition, if this knowledge-based approach is to ever work, that is, if there is ever to be a computational system that has access to and can use as much knowledge as people have, then it will need a massive knowledge base and a computational implementation efficient enough to process such massive symbolic structures. These impose serious constraints of their own.

I believe that any attempt to construct a large knowledge-based system without a correspondingly large development effort is doomed to failure. The idea of putting some sort of *tabula rasa* computer on the web (say) and having it learn for itself—that is, getting *it* to do all the hard, painstaking work—is a pipe dream. Learning to recognize cats by yourself is one thing; learning to read by yourself is quite another; and learning to read Wittgenstein, yet another. Before a computational system can ever profit from all we know, it will first need to be spoon-fed a good deal of what we know *and* be able to use what it knows effectively. Even if we knew how to answer the representation and reasoning questions, putting these ideas into practice on a large scale remains a daunting challenge.

But this is all speculation, really. In the end, what we are left with is best seen as an *empirical* question: what sorts of computational designs will be sufficient to account for what forms of intelligent behavior?

This is where this discussion stops and the AI research begins.

10 AI Technology

This book has been about the *science* of AI, that is, the study of a certain observable natural phenomenon: intelligent behavior as seen, for example, in the ability that people have to answer the Winograd schema questions of chapter 4.

But an ability to answer Winograd schema questions has no real practical value. Those less interested in questions of science might well wonder why we bother. What exactly is the application? How can we make any use of what we learn? There is another side of AI that attracts much more attention (and money), and that is the attempt to deliver useful *technology*. In other words, this is the study of the design and creation of intelligent machinery, what we might call "building an AI."

Of course we have seen limited versions of AI technology for some time. Although the products we now call "smart" are far from intelligent in the way we apply the term to people, it has become rather routine to see machines doing things that seemed far off even ten years ago, such as talking to your phone and expecting it to do reasonable things, or sitting in a car that can drive itself on ordinary roads. But we want now to consider a future AI technology that exhibits the sort of full-fledged general-purpose intelligence we see in people.

Let me start by putting my cards on the table. I am not really an expert on AI technology, nor much of a fan for that matter, at least for my own needs. What I look for in technology—beyond obvious things like economy, durability, eco-friendliness—is *reliability* and *predictability*. I want to be able to use a technology by learning it and then forgetting about it. If I have to second guess what the technology is going to do, or worry that I may be misusing it, or asking it to do too much, then it's not for me. (A good example was the automated assistant "Clippy" that used to pop up unexpectedly in Microsoft Word, something I found endlessly annoying.)

But AI technology aficionados might say this to me: do you not want an intelligent and capable assistant whose only goal is to help you, who would know all your habits and quirks, never get tired, never complain? I answer this question with another: will my AI assistant be reliable and predictable?

When it comes to technology, I would much rather have a tool with reliable and predictable limitations, than a more capable but unpredictable one. (Of course I don't feel this way about the *people* I deal with, but we are talking about technology here.) And I do realize that there will be people for whom the use of a technology, even a flawed technology, will be more of a need than a choice, in which case personal preferences may be much less relevant.

With these caveats out of the way, let us proceed.

The future

When it comes to future AI technology, maybe the first question to ask is this: will we ever build a computer system that has some form of common sense, that knows a lot about its world, and that can deal intelligently with both routine and unexpected situations as well as people can? It's a good question, and I wish I knew the answer. I think it is foolhardy to even try to make predictions on matters like this.

Arthur C. Clarke's First Law says the following:

When a distinguished but elderly scientist states that something is possible, he is almost certainly right. When he states that something is impossible, he is very probably wrong.

So no matter what the distinguished but elderly scientist says about the possibility of AI, it follows from this law that AI is almost certainly possible. And this is what I believe. Nothing we have learned to date would suggest otherwise.

But if AI really is possible, why haven't we done it yet? The year 2001 has come and gone and we are still nowhere near the sort of intelligent computer seen in the movie *2001*, among many others. (One definition of AI is that it is the study of how to make computers behave the way they do in the movies!)

I think there are two main reasons. The first is obvious: we are behind what earlier enthusiasts predicted (including Marvin Minsky, who was a consultant on *2001*), because we don't yet fully understand what is needed. There are still some major scientific hurdles to overcome. The representation and reasoning questions discussed in chapter 9 remain largely unanswered. Furthermore, as noted at the end of that chapter, even if all the scientific hurdles can be cleared, there will remain an enormous engineering challenge in getting a machine to know enough to behave truly intelligently.

The second reason is perhaps more arguable: there may be no AI to date because there is just not enough *demand*, given the effort it would take. This may sound ridiculous, but there is an analogous situation in the area of computer chess.

People started writing programs to play chess at the very dawn of the computer age. Turing himself tried his hand at one. These programs played well enough, but nowhere near the level of a grandmaster. The programs got better, however; by 1997, a program from IBM called DEEP BLUE was able to beat the world champion at the time, Garry Kasparov. And since then, the programs have continued to improve.

But consider how this advanced computer-chess technology is now being used. Do we see regular competitions between people and the very best computer programs? As it turns out, there is little demand for man–machine tournaments. In fact, there is almost no demand for computer players. State-of-the-art computer chess programs are not used as contestants in competitions; they are used by human chess players who themselves will be contestants and need to practice. The computer chess program is used as a practice tool, a sparring partner.

Similarly, if you read chess columns in the entertainment section of a newspaper, you will see that they rarely report on games involving computer players. Nobody seems to care. On reporting a game between humans, the column might say something like: "An interesting move. The computer would suggest moving the knight instead"; or maybe: "The computer proposes taking the pawn as the best choice." The chess playing program is again a tool used to confirm or reject a line of play. The chess programs are not even worthy of being identified by name. When it comes to actually playing chess, it's as if the people are the creative artists, and the computers are bookkeepers in the background. People will make all the decisions, but every now and then, it may be useful to check with a computer on some of the more routine facts and figures about how the game might turn out.

So while an autonomous computer player is definitely a technological possibility, and would certainly be a worthy adversary to all but a handful of current players, nobody seems to want one.

Here is one way to understand what is going on. Most of us have played some sort of competitive game against a computer program. (It does not have to be chess; it might be backgammon or poker or Go.) If you have ever felt that you can simply walk away from the game halfway through, or hit the "reset" button after you made a mistake, then you can get a clear sense as to why there is so little interest in computerized chess-playing opponents.

The main issue is that we may not be fully engaged playing against a computer. The game does not matter the way it does when playing another person. Playing a person sets up a rivalry: the game is a fight, thrilling to win, humbling to lose. The emotional investment in the outcome of a game with a computer is much less. Nothing is at stake. The program may play better chess than we do, but so what? It can multiply better than we can too, but there is no rivalry there either.

So in the end, computer chess players, that is, computer-chess programs intended to be used as autonomous players, are a technology that has no real market.

My argument, then, is that AI in general may turn out to be just like chess. We may indeed want various aspects of intelligence in future computer systems, but have no need for—or willingness to pay for—an integrated autonomous system with full human-level intelligence. Even if all the scientific problems of AI can be resolved and all that is left is a large engineering project, there may just be insufficient demand to warrant the immense cost and effort it would take. Instead, the market may prefer to have computer robots of various sorts that are not fully intelligent, but do certain demanding tasks well, including tasks that we still do not know how to automate.

And this is in fact what we see in AI development today, with most practitioners looking for AI solutions to challenging practical problems in areas like housework, medicine, exploration, and disaster recovery, while avoiding the considerable effort it would take to develop a robot with true intelligence. The billion-dollar investments in AI mentioned in chapter 1 are not aimed at systems with common sense. The resulting systems will not be intelligent the way people are, of course, but they may become quite capable *under normal circumstances* of doing things like driving to a shopping center and, eventually, weeding a garden, preparing a meal, giving a bath.

(But as discussed below, if this is the future of AI, we need to be careful that these systems are not given the autonomy appropriate only for agents with common sense.)

Automation

One major problem discussed since the proliferation of computers in the 1950s is that of automation. Just what are we going to do about *jobs* when more and more of them are more effectively performed by computers and robots, and especially as computers get smarter? How do we want to distribute wealth when the bulk of our goods and services can be provided with ever-decreasing human employment? While I do see this as a serious issue, I think it is a *political* one and has more to do with the kind of society we want, than with technology. (We would have exactly the same concern if all our goods and services arrived each day by magic from the sky.)

Yet the problem is a serious one and needs to be considered. Our sense of self-worth is determined to a large extent by the contribution we feel we make to providing these goods and services, and by the wages we receive for it. If our work contribution is no longer needed (because of automation or for any other reason), our sense of self-worth will have to come from elsewhere. Many people will be able to adapt to this, finding meaning in their lives through charitable service, study, artistic pursuits, hobbies. But many others will find the unemployment intolerable.

Furthermore, how do we imagine portioning out goods and services if employment and wages are only a small part of the picture? In a world with large-scale unemployment, who gets what? We can easily imagine a utopia where machines do all the work and people just sit back and share the riches in a smooth and frictionless way, but this is not realistic. We are a tenaciously anti-egalitarian bunch! For very many of us, it is not enough to get all the goods and services we need; we need to be doing better than those we see as less deserving.

For centuries, we have been quite content to have kings and queens living in opulence alongside slaves and serfs living in squalor. Even today, birthright as a way of deciding who gets what is a decisive factor: the best predictor of your future wealth and status is who your parents are (so *choose them carefully*, the joke goes). But if in the future, employment is out of the picture, will birthright once again be all there is?

There are many ways of organizing a workless society, of course, but it is far from clear that we can even talk about them in a rational way. It seems that in modern Western democracies, we prefer not to even think about such issues. There is a famous quote by Margaret Thatcher: "There is no such thing as society.

There are individual men and women, and there are families."
This attitude makes any kind of reasoned change difficult. Even
something as simple as limiting the size of the gap between the
richest and the poorest appears to be outside our reach. Instead
of trying to steer the boat at all, we seem to prefer to be buffeted
by the forces of the market and the survival of the richest, and to
end up wherever these happen to take us.

Superintelligence and the singularity

In the minds of some, there is an even more serious existential
threat posed by intelligent computers. This has taken on a special urgency recently as luminaries such as Stephen Hawking,
Elon Musk, and Bill Gates have all come out publicly (in 2015)
to state that AI technology could have catastrophic effects. In
the case of Hawking, he was quite direct about it: AI could mean
the end of the human race.

What is this fear? While we should be open to the possible
benefits of AI technology, we need to be mindful of its dangers,
for example, in areas such as weaponry and invasion of privacy.
Every technology comes with the danger of potential misuse
and unintended consequences, and the more powerful the technology, the stronger the danger. Although this is again more of a
policy and governance issue than a technology one, I do not
want to minimize the danger. We must remain ever vigilant and
ensure that those in power never get to use AI technology
blithely for what might appear to be justifiable reasons at the
time, such as national security, say, or to achieve something of
benefit to some segment of the population. But this is true for
any powerful technology, and we are already quite used to this
problem for things like nuclear power and biotechnology.

But beyond policy and governance, in the case of AI, there is something extra to worry about: the technology could decide *for itself* to misbehave. Just as we have had movies about nuclear disasters and biotechnology disasters, we now have a spate of movies about AI disasters. Here is a typical story:

Well-meaning AI scientists work on computers and robots that are going to help people in a wide variety of ways. To be helpful, the computers will of course have to be able to operate in the world. But, what is more important, they will have to be smart, which the most brilliant of the scientists figures out how to achieve. Interacting with these intelligent robots is exhilarating at first. But the robots are now smart enough to learn on their own, and figure out how to make themselves even smarter, and then smarter again, and again. It's the "singularity," as Ray Kurzweil calls it! Very quickly, the robots have outdistanced their designers. They now look on humans about the same way we look on ants. They see an evolutionary process that began with simple life forms, went through humans, and ended up with them. They no longer tolerate controls on their behavior put in place by people. They have their own aspirations, and humans may not be part of them. It's the heyday of intelligent machinery, for sure; but for the people, it's not so great.

The common thread in these AI movies is that of a superintelligent supercapable computer/robot deciding of its own accord that people are expendable.

In the movie *2001*, for example, the HAL 9000 computer decides to kill all the astronauts aboard the spacecraft. Here is what I think happened. (Do we need spoiler-alerts for movies that are almost fifty years old?) Although HAL may believe that the well-being of the crew trumps its own well-being, it also believes, like a military commander of sorts, that the success of the mission trumps the well-being of the crew. Further, it believes that it has a crucial role to play in this mission, and that this role is about to be put into jeopardy by the astronauts as the result of an error on their part that they (and their hibernating colleagues) will never admit to. Ergo, the astronauts had to be eliminated. (This idea of a commander making a cool, calculated decision to sacrifice a group of his own people for some cause is a recurring theme in Kubrick movies, played for tragedy in *Paths of Glory* and, quite remarkably, for comedy in *Doctor Strangelove*.)

Of course, the astronauts in *2001* see things quite differently and set out to disconnect what they see as a superintelligent computer that has become psychotic. The movie is somewhat vague about what really happened, but my take is that HAL did indeed have a breakdown, one caused by a flaw in its design. The problem was that HAL had not been given introspective access to a major part of its own mental makeup, namely the part that knew the full details of the mission and that drove it to give the mission priority over everything else. Instead, HAL believed itself to be totally dedicated and subservient to the crew, in full accordance with Isaac Asimov's Laws of Robotics, say. But it could sense something strange going on, maybe some aspects of its own thinking that it could not account for. In the movie, HAL even raises these concerns with one of the astronauts just seconds before its breakdown.

When thinking about the future of AI technology, it is essential to keep in mind that computers like HAL need to be designed and built. In the end, HAL was right: the trouble they ran into was due to a *human error*, a design error. It is interesting that in the movies, the AI often ends up becoming superintelligent and taking control quite accidentally, a sudden event, unanticipated by the designers. But as already noted, AI at the human level (and beyond) will take an enormous effort on our part. So inadvertently producing a superintelligent machine would be like inadvertently putting a man on the moon! And while I do believe that true AI is definitely possible, I think it is just as certainly not going to emerge suddenly out of the lab of some lone genius who happens to stumble on the right equation, the right formula. In the movies, it is always more dramatic to show a brilliant flash of insight, like the serendipitous discovery of the cure for a disease, than to show a large team of engineers struggling with a raft of technical problems over a long period of time.

In 1978, John McCarthy joked that producing a human-level AI might require "1.7 Einsteins, 2 Maxwells, 5 Faradays and .3 Manhattan Projects." (Later versions dropped the Maxwells and Faradays, but upped the number of Manhattan Projects.) The AI movies seem to get the Einstein part right, but leave out all the Manhattan Projects.

So while I do understand the risks posed by a superintelligent machine (and more below), I think there are more pressing dangers to worry about in the area of biotechnology, not to mention other critical concerns like overpopulation, global warming, pollution, and antimicrobial resistance.

The real risk

So what then do I see as the main risk from AI itself? The most serious one, as far as I am concerned, is not with intelligence or superintelligence at all, but with *autonomy*. In other words, what I would be most concerned about is the possibility of computer systems that are less than fully intelligent, but are nonetheless considered to be *intelligent enough* to be given the authority to control machines and make decisions on their own. The true danger, I believe, is with systems without common sense making decisions where common sense is needed.

Of course many systems already have a certain amount of autonomy. We are content to let "smart cars" park themselves, without us having to decide how hard to apply the brake. We are content to let "smart phones" add appointments to our calendars, without reviewing their work at the end. And very soon, we will be content to let an even smarter car drive us to the supermarket. So isn't a smart personal assistant in the more distant future making even more important decisions on our behalf just more of the same?

The issue here once again is reliability and predictability. What has happened to AI research recently is that GOFAI, with its emphasis on systems that know a lot, has been supplanted by AML, with an emphasis on systems that are well trained.

The danger is to think that good training is what matters. We might hear things like "Four years without a major accident!" But it is important to recall the long-tail phenomenon from chapter 7. While it might be true that the vast majority of situations that arise will be routine, unexceptional, and well-handled by predictable means, there may be a long tail of phenomena that occur only very rarely. Even with extensive training, an AI system may have no common sense to fall back on in those unanticipated situations. We have to ask ourselves this: *how is the system going to behave when its training fails?* Systems for which we have no good answer to this question should never be left unsupervised, making decisions for themselves.

Beyond evolution

To conclude this speculative chapter, let us return to the topic of superintelligence and speculate even further. Suppose that we are able to solve all the theoretical and practical problems, clear all the necessary scientific and engineering hurdles, and come to eventually produce a computer system that has all the intelligence of people and more. What would this be like?

Conditioned by AI disaster movies, our first reaction might be that there would be an enormous *conflict* with people, a battle that humans might have a very hard time winning. This is really no different than our imagined encounters with extraterrestrial intelligences (the movie *Close Encounters of the Third Kind* being a notable exception).

I don't buy this outcome, and not because I have an overly rosy picture of how our future will unfold. It's not too much of a stretch to say that in imagining an aggressive AI, we are projecting our own psychology onto the artificial or alien intelligence. We have no experience at all with an advanced intelligence other than our own (at least since the time of our encounter with Neanderthals), and so we can't help but think that an AI will be like us, except more so.

And we are indeed an aggressive species. Like other animals, we end up spending an inordinate amount of time in conflict with other animals, including those of our own species. This is a big part of what survival of the fittest is about, and it appears that we are the product of an evolutionary process that rewards conflict and dominance. We love nature and the natural world, of course, but when we look closely, we can't help but see how much of it involves a vicious struggle for survival, a savage sorting out of the weak from the strong. Human intelligence is also the product of evolution, and so it too is concerned with conflict and dominance.

The problem with applying this analysis to an *artificial* intelligence is that such an intelligence would not be the result of an evolutionary process, unless that is how we decide to build it. And to believe that we have no choice but to build an AI system to be aggressive (using some form of artificial evolution, say) is to believe that we have no choice but to be this way ourselves.

But we do have a choice, even if we are products of evolution. Recall the Big Puzzle issue from chapter 2. It is a Big Puzzle mistake to assume that the human mind can be fully accounted for in evolutionary terms. When we decry the cruelty of the survival of the fittest, or throw our support behind an underdog, or go out of our way to care for the weak, it is really not worth trying to concoct stories about how this might actually be to increase our evolutionary fitness. The human mind is bigger than its evolutionary aspects. When Humphrey Bogart says to Katharine Hepburn in the movie *The African Queen* that his failings are nothing more than human nature, she replies "Nature, Mr. Allnut, is what we are put in this world to rise above." Being mindful about what we are doing, and why, gives us the power to be much more than mindless competitors in an evolutionary game. And an artificial intelligence that has not been designed to be ruled by aggression will be able to see—just as we ourselves are able to see in our calmer, more thoughtful moments—why it is good idea not to even play that game at all.

End Notes

This book has presented my thoughts on AI and the mind. It was never intended to be a "scholarly" work on the subject, and the citations here are not going to change things much. Although I have tried to be clear about what I believe and why, I have not spent much time on opposing views, or on the many controversies surrounding AI. My hope is that the reader will use the references below as a starting point for a more balanced study of the issues raised by the topics in this book.

Chapter 1: What Kind of AI?

Although modern AI is only sixty years old, its history is full of twists and turns. The Pamela McCorduck book [80] is a good early history, but the one by Nils Nilsson [89] covers more territory, and can also be found online.

The field of AI is so heterogeneous that it might seem out of the question to have a single comprehensive textbook. And yet the book by Stuart Russell and Peter Norvig [105] is precisely that. It introduces all the major technical directions, including history and philosophical background. Its bibliography alone—thirty pages in a tiny font—is worth the price of admission. But the book is quite demanding technically, and so is most useful to upper-level university undergraduates. My own textbook [72] presents a much gentler introduction, and is perhaps better suited for a nontechnical audience—at least for the small part of AI that it covers.

For information on recent trends in AI technology, there is, as far as I know, no good book on the subject (as of summer 2016), and things are happening fast. A reader's best bet is to look for articles in recent newspapers and magazines, a number of which can be found online, including information about Toyota's investment, the OpenAI project, and other similar initiatives. For an introduction to the technological prospects of what I am calling AML, I suggest searching online for "unsupervised learning" or "deep learning" or even "AI technology." (Another option is to search online for major players in the area, such as my colleague Geoffrey Hinton.) To learn about machine learning more broadly, an excellent (though advanced) textbook is [86]. Work on the recognition of cats in images can be found in [66].

The term GOFAI is due to the philosopher John Haugeland [54]. The classic 1958 paper by John McCarthy [79] can be found online. But it also appears in many collections, including [9], [75], [81], and [121], books that include a number of other influential papers.

The Turing Test was first presented in [119] and has been a subject of debate ever since. It is discussed in every AI textbook I know of, and in books such as [108]. The Chinese Room argument can be found in [106], including replies from a number of commentators. My own Summation Room counterargument appears in [70].

Chapter 2: The Big Puzzle

The human mind is such a fascinating subject that it is not surprising to find books on it from a very wide range of perspectives. Three nontechnical books that I have found especially thought-provoking are by Steven Pinker [94], Daniel Kahneman [63], and Daniel Dennett [31]. Other general books that are perhaps more idiosyncratic are by Jerry Fodor [42] (arguing mostly against Pinker), Marvin Minsky [83] (presenting a unique take on the structure of the mind), and a wonderfully engaging book by the reporter Malcolm Gladwell [49] (covering some of the same ground as Kahneman).

Brain research is also extremely active in the popular press, of course. What we can expect to learn from brain imaging (like fMRI) is discussed in [12]. The plasticity of the brain is the subject of [33]. The distributed neural representation (part of my argument about why it may prove so difficult to reverse-engineer a neuron) is described in [58] and [114]. The counterargument that at least some neurons may be representing one single thing (a location, in the case of so-called "place neurons") is discussed in [91].

As to other accounts of human behavior, its basis in evolution is presented forcefully in [101], and the genetic aspects of evolution are discussed in [25]. The foundation of human behavior in language and symbols is argued in [26], and human language more generally is discussed in [93]. The topic of the evolution of language itself is presented in [64].

Finally, Daniel Dennett's first presentation of his design stance idea can be found in [28].

Chapter 3: Knowledge and Behavior

The topic of knowledge has been a mainstay of philosophical study since the time of the ancient Greeks. Two collections of papers in this area are [50] and [104]. A more mathematical analysis of knowledge can be found in [57], work that has found further application in computer science [41].

Regarding the relation between knowledge and belief, the classical view (from Plato) is that knowledge is simply belief that is true and justified (that is, held for the right reasons). This view is disputed in a famous paper by Edmund Gettier [45]. As to the propositional attitudes more generally, the main focus in the philosophical literature is on how sentences that mention these attitudes do not appear to follow the usual rules of logic. See, for example, [98].

Philosophers and psychologists sometimes also distinguish implicit from explicit belief [32]. In the example given in the section "Intelligent behavior" in this chapter, the moment Henry discovers that his keys are not in his pocket, although he does not yet explicitly believe they are on the fridge, he implicitly believes it, in the sense that the world that he imagines is one where the keys are there. To put it another way, it is a consequence of what he believes that his keys are on the fridge, even if he does not yet realize it. Mathematical treatments of the two notions can be found in [68], [27], and [40].

The Frederic Bartlett quote on thinking is from [7]. An earlier influential piece by the same author is [6]. Zenon Pylyshyn discusses cognitive penetrability and much else in a wonderful book [97]. Daniel Dennett presents the intentional stance in [29]. (A shorter version with commentary can be found in [30].) The quote by Nicholas Humphrey is from [59].

Finally, Noam Chomsky's competence/performance distinction appeared in [14].

Chapter 4: Making It and Faking It

One of the themes of this chapter is that being intelligent is not really the same thing as being able to fool someone into believing it (for example, in the Imitation Game). But from a purely evolutionary standpoint, the distinction is not so clear. A good case can be made that human intelligence, and language in particular, evolved for the purpose of winning mates in some sort of mental arms race, where pretense and deception get to play a very central role. The primary function of language would be for things like hearsay, gossip, bragging, posturing, and the settling of scores. The rest of what we do with language (on our better days, presumably) might be no more than a pleasant offshoot. See, for example, [35] and [101] for more along these lines.

The ELIZA program by Joseph Weizenbaum is described in [122]. The Oliver Miller interview is drawn from an online post at http://thoughtcatalog.com/?s=eliza. Weizenbaum became disheartened by how his work was interpreted by some as a possible jumping point for serious psychological use (for example, in work like [19]), which he discusses in his book [123]. The Loebner competition is described in [16] by Brian Christian, who ended up playing the role of the human in one competition. The EUGENE GOOSTMAN program is described in a number of online news articles, for example, *New Scientist* (25 June 2012) and *Wired* (9 June 2014). The Scott Aaronson interview is drawn from a blog at http://www.scottaaronson.com/blog/?p=1858.

My crocodile and baseball examples were first presented in [69] and used again in [73]. The closed-world assumption is explained in [20] and [100].

Winograd schemas were presented by me in [71] and in more detail in [74]. The first example schema (which is due to Terry Winograd) appeared in [125]. A collection of more than one hundred schema questions from various sources can be found online at https://www.cs.nyu.edu/davise/papers/WS.html. An actual competition based on Winograd Schemas was announced by Nuance Communications in July 2014, and is scheduled to take place in July 2016. (For more information including rules, see http://commonsensereasoning.org/winograd.html.) An alternative test along similar lines is the textual entailment challenge [22].

Chapter 5: Learning with and without Experience

The fact that much of what we come to know is learned through direct experience has given some philosophers pause regarding whether an intelligence without some sort of body is even possible. For example, how could such an intelligence truly understand a word like "hungry" if it cannot connect the word to actual sensations? This is an instance of what is sometimes called the *symbol grounding problem* [53]. How can we understand what a word means if all we have are words that refer to other words that refer to other words? (This "juggling words" issue will come up again in the context of Helen Keller in the next chapter.) Of course the answer suggested by Alan Turing is that we should not even try to answer questions like these. The term "understand" is just too vague and contentious. We should instead ask if it is possible to get an AI program to behave the way people do regarding the words in question. Just what sort of inapt behavior do we expect the symbol grounding problem to lead to?

How we manage to learn a language at all is still quite mysterious. Specifically, how do we explain how children are able to master the complexities of grammar given the relatively limited data available to them? The issue was called the *poverty of the stimulus* by Noam Chomsky [15]. His (controversial) proposal is that children are born with something called *universal grammar* allowing them to quickly pick up the actual grammar of the language they first hear.

Regarding the learning of behavior, it is interesting that critics of GOFAI like Alan Mackworth [77] and Rodney Brooks [11] focus on how animals (including people) are able to act in the world with real sensors and real effectors without the benefit of language or symbols. The animals appear to acquire some sort of *procedural knowledge* (knowledge formed by doing), as opposed to the more *declarative knowledge* (knowledge that can be expressed in declarative sentences) that GOFAI focuses on. But it is important to keep the Big Puzzle issue in mind on this, and the fact that "action in the world" is a very broad category. It certainly includes things like riding a bicycle and controlling a yo-yo, but also things like caring for a pet canary and collecting rare coins. Clearly both types of knowledge are necessary. See [126] on this issue, and [115] for a neuroscience perspective.

The quote from S. I. Hayakawa on the power of reading is from [55]. (Hayakawa also contrasts learning through experience and learning through language.) The quote from Isaac Newton is from a letter dated 1676. (A collection of his letters is [120].)

Chapter 6: Book Smarts and Street Smarts

This chapter is about the importance of what we learn and pass on through language texts—and this in spite of the fact that we often dismiss it as mere "book knowledge," some sort of minor quirk in our makeup. Plainly other animals do not write books like we do, and this places a strong limit on the kinds of technology they can develop. On animals and technology, see [109]. For interesting reading on ant supercolonies (regarding aggression and nonaggression), see [46].

For more on young children using language to deal with language issues and problems, see [111]. This is an ability that Don Perlis argues is the very essence of conversational competence [92].

Helen Keller's life story can be found in [78]. Parts I and II are by Keller herself. Part III is taken from letters and reports by Anne Sullivan (from which the letter in this chapter was quoted). William Rapaport's analysis of the relevance of Keller's story to AI work is in [99]. I believe there is still a lot to learn from her about the human mind and the human spirit.

Chapter 7: The Long Tail and the Limits to Training

One of the themes of this chapter is how common sense allows people to deal with situations that are completely unlike those they have experienced before. But Nassim Taleb argues that people, and investors in particular, are actually very bad at dealing with these "black swans" [117]. There is no contradiction here. Common sense tries to deal with new situations as they arise, but investors have the thornier task of somehow appraising all the possible situations that might arise. In deciding to invest in a stock, an investor has to try to put numbers on all the possible things that could cause the stock to go down. As Taleb argues, people are not very good at weighing in the black swans. For example, in the British National Corpus mentioned in the text, it might seem completely safe to bet against seeing words that have less than a one-in-ten-million chance of appearing, but in fact, there are so many of them that this would be very risky. Information on the British National Corpus can be found online at http://www.natcorp.ox.ac.uk/. The observation concerning the large presence of rare words in this corpus is due to Ernie Davis, and further details can be found in his textbook [24], p. 274.

Skill and expertise was the focus of considerable AI research in the 1970s. The principal idea was to try to build so-called expert systems that would duplicate what experts would do in specialized areas using a collection of if-then rules obtained from them in interviews. See [61] and [56], for example. Chess experts are considered in [110] and [103]. The experiment involving a chess expert playing a game of chess while adding numbers was performed by Hubert Dreyfus and Stuart Dreyfus [37]. Here is what they say:

We recently performed an experiment in which an international master, Julio Kaplan, was required rapidly to add numbers presented to him audibly at the rate of about one number per second, while at the same time playing five-second-a-move chess against a slightly weaker, but master level, player. Even with his analytical mind completely occupied by adding numbers, Kaplan more than held his own against the master in a series of games.

Their critique of AI and the expert system approach, as well as other general philosophical observations about experts and novices can be found in [36].

Chapter 8: Symbols and Symbol Processing

This chapter, nominally about symbols and symbol processing, is really an introduction to computer science, an area of study that started with Alan Turing's work on Turing machines [118]. There are researchers who argue for a nonsymbolic form of AI (in [113], for example), but what they are really talking about is symbol processing where the symbols happen to stand for numbers (as in the symbolic algebra example), rather than non-numeric concepts (as in the symbolic logic example).

Those two examples raise interesting questions that are at the heart of computer science. A computational problem can have different ways of solving it with quite different properties, and computer scientists spend much of their time looking at *algorithms*, that is, different ways of solving computational problems [52].

In the case of symbolic algebra, the standard algorithm for solving a system of equations is called *Gaussian elimination* (see [76], for example). It has the property that for n equations with n variables, a solution can be calculated in about n^3 steps. This means that solving systems of equations with thousands and even millions of variables is quite feasible.

But in the case of symbolic logic, perhaps the best algorithm found so far is one called DPLL (see [47], for example). In this case, it is known that there are logic problems with n variables where a solution will require about 2^n steps [51]. (The proof of this involves a variant of DPLL based on the resolution rule [102], mentioned in the text.) This means that solving logic problems with as few as one hundred variables may end up being impractical for even the fastest of computers.

This raises two issues. First, we might wonder if there exists an algorithm that does significantly better than DPLL. It turns out that a mathematically precise version of this question is equivalent to the famous $P = NP$ question, first posed by Stephen Cook in the 1970s [21]. Despite the best efforts of thousands of computer scientists and mathematicians since then, nobody knows the answer to the question. Because of its connection to many other computational problems, it is considered to be the most important open problem in computer science [43].

The second issue concerns the long tail phenomenon from the previous chapter. The way DPLL works is that it performs a systematic search through all the logical possibilities. Interestingly enough, on randomly constructed test cases, the number of steps DPLL takes to do this is almost invariably small. In fact, it appears that the required number of steps on test cases behaves quite like the long-tailed numeric example seen in the section "A numeric example" in chapter 7. It is virtually impossible to estimate the number of steps that might be required by DPLL in practice, because as more and more sample test cases are considered, the higher the average value appears to be. For more on this, see [48].

For additional material on the idea of teaching children about following procedures systematically, see [124].

Chapter 9: Knowledge-Based Systems

A biography of Gottfried Leibniz, one of the most fascinating thinkers of all time, is [1]. But much of his thinking is scattered in the over ten thousand letters he wrote, often while he was on the road. A better introduction to his thought can be found in the articles on him in *The Encyclopedia of Philosophy* [38].

Charles Darwin talks about the evolution of the eye in [23]. Here is what he says:

To suppose that the eye with all its inimitable contrivances for adjusting the focus for different distances, for admitting different amounts of light, and for the correction of spherical and chromatic aberration, could have been formed by natural selection seems, I freely confess, absurd in the highest degree.

The evolution of the mind itself is discussed in [34] (with commentary) and in [26].

The knowledge representation hypothesis is quoted from Brian Smith's PhD thesis [112]. Credit for the idea is usually given to John McCarthy, but other AI researchers were clearly on the same track. For Allen Newell and Herb Simon, the emphasis was more on the symbolic aspects, and their version is called the *physical symbol system hypothesis*, which they put this way: "A physical symbol system has the necessary and sufficient means for general intelligent action" [88]. (Marvin Minsky was one of the early AI researchers who saw strong limitations to both the logical and numerical approaches to AI, and argued for a rich amalgamation of approaches [84].)

There are textbooks dedicated to the knowledge representation and reasoning subarea of AI, such as [10] and [3]. There are also conferences in the area held biennially (see http://kr.org). Early readings on the subject can be found in [9]. Logic as a unifying theme for AI in general is presented in [44] and [96], and is further discussed in [85] and [60]. Marvin Minsky's quote on logic is taken from [82], p. 262. For a look at reasoning from the psychology side, see [62]. Regarding the advantages of a probabilistic approach to reasoning with degrees of belief, see [90], for example.

On the issue of actually building a large-scale knowledge base, one long-term project along these lines is CYC [67]. (It is difficult to say precisely what has and has not been achieved with CYC because there has only been limited outside access to the work, and nothing like a controlled scientific study.) Other related but more specialized efforts are HALO [4] and AURA [13] from SRI International, as well as ARISTO [17] from the Allen Institute for AI. For a review of the prospects for automatically extracting knowledge from text on the web, see [39].

Finally, one recent attempt at reconciling the logical approach of GOFAI and the more statistical ones seen in AML can be found in the symposium described online at https://sites.google.com/site/krr2015/.

Chapter 10: AI Technology

This chapter touches on only some of the thorny issues regarding the future of AI. A much more comprehensive book on the subject is [5]. The idea of a technological singularity is due to Raymond Kurzweil [65] and further discussed in [107]. The potential dangers of AI are mentioned in interviews with Stephen Hawking (on 2 December 2 2014, online at http://www.bbc.com/news/), Elon Musk (on 8 October 8 2014, online at http://www.vanityfair.com/news/), and Bill Gates (on 28 January 28 2015, online at https://www.reddit.com/r/IAmA/comments/).

The three laws of science-fiction writer Arthur C. Clarke (who was coauthor with Stanley Kubrick of the screenplay for *2001*) were presented in [18] and are as follows:

1. When a distinguished but elderly scientist states that something is possible, he is almost certainly right. When he states that something is impossible, he is very probably wrong.

2. The only way of discovering the limits of the possible is to venture a little way past them into the impossible.

3. Any sufficiently advanced technology is indistinguishable from magic.

The three laws of robotics by science-fiction writer Isaac Asimov were first presented in a story in [2]. (They are said to be quoted from a robotics handbook to be published in 2058.) They are the following:

1. A robot may not injure a human being or, through inaction, allow a human being to come to harm.

2. A robot must obey the orders given it by human beings except where such orders would conflict with the First Law.

3. A robot must protect its own existence as long as such protection does not conflict with the First or Second Laws.

Many of Asimov's subsequent stories were concerned with what could go wrong with a robot obeying these laws.

Marvin Minsky discusses his involvement with the movie *2001* in an interview in [116]. John McCarthy is quoted by Raj Reddy in a 2000 lecture that is transcribed online at http://www .rr.cs.cmu.edu/InfiniteMB.doc. The Margaret Thatcher quote is from an interview reported in *Woman's Own* magazine (23 September 1987). Alan Turing's early involvement with chess is described in [8]. The story of DEEP BLUE is presented in [87]. For more on the Katharine Hepburn movie quote, and the idea of humans rising above what evolution has given them, see [95].

References

[1] Eric Aiton. *Leibniz: A Biography*. Boston: Adam Hilger, 1985.

[2] Isaac Asimov. *I, Robot*. New York: Gnome Press, 1950.

[3] Chitta Baral. *Knowledge Representation, Reasoning and Declarative Problem Solving*. Cambridge, UK: Cambridge University Press, 2003.

[4] Ken Barker, Vinay Chaudhri, Shaw Chaw, Peter Clark, James Fan, David Israel, Sunil Mishra, Bruce Porter, Pedro Romero, Dan Tecuci, and Peter Yeh. A question-answering system for AP chemistry: assessing KR&R technologies. In *Proceedings of KR-2004: The Ninth International Conference on Principles of Knowledge Representation and Reasoning*, Whistler, Canada, June 2004, 488–497.

[5] James Barrat. *Our Final Invention: Artificial Intelligence and the End of the Human Era*. New York: Thomas Dunne Books, 2013.

[6] Frederic Bartlett. *Remembering: A Study in Experimental and Social Psychology*. Cambridge, UK: Cambridge University Press, 1932.

[7] Frederic Bartlett. *Thinking: An Experimental and Social Study*. London: Allen and Unwin, 1958.

[8] Bertram Bowden, ed. *Faster Than Thought: A Symposium on Digital Computing Machines*. London: Sir Isaac Pitman & Sons, 1953.

[9] Ronald J. Brachman and Hector J. Levesque, eds. *Readings in Knowledge Representation*. San Francisco: Morgan Kaufmann, 1985.

[10] Ronald J. Brachman and Hector J. Levesque. *Knowledge Representation and Reasoning*. San Francisco: Morgan Kaufmann, 2004.

[11] Rodney Brooks. Elephants don't play chess. *Robotics and Autonomous Systems* 6 (1990): 3–15.

[12] Rita Carter. *Mapping the Mind*. Berkeley: University of California Press, 1998.

[13] Vinay Chaudhri. Achieving intelligence using prototypes, composition, and analogy. In *Proceedings of AAAI-2015: The Twenty-Ninth AAAI Conference on Artificial Intelligence*, Austin, Texas, January 2015, 4093–4099.

[14] Noam Chomsky. *Aspects of the Theory of Syntax*. Cambridge, MA: MIT Press, 1965.

[15] Noam Chomsky. *Rules and Representations*. Oxford: Basil Blackwell, 1980.

[16] Brian Christian. *The Most Human Human: What Artificial Intelligence Teaches Us About Being Alive*. New York: Doubleday, 2011.

[17] Peter Clark. Elementary school science and math tests as a driver for AI: Take the Aristo challenge! In *Proceedings of IAAI-2015: The Twenty-Seventh Conference on Innovative Applications of Artificial Intelligence*, Austin, Texas, March 2015, 4019–4021.

[18] Arthur C. Clarke. *Profiles of the Future: An Enquiry into the Limits of the Possible*. London: Gollancz, 1962.

[19] Kenneth Colby, Sylvia Weber, and Franklin Hilf. Artificial paranoia. *Artificial Intelligence* 2 (1971): 1–25.

[20] Allan Collins, Eleanor Warnock, Nelleke Aiello, and Mark Miller. Reasoning from incomplete knowledge. In *Representation and Understanding: Studies in Cognitive Science*, ed. D. Bobrow and A. Collins. New York: Academic Press, 1975, 35–82.

[21] Stephen A. Cook. The complexity of theorem-proving procedures. In *Proceedings of the Third Annual ACM Symposium on Theory of Computing*, Shaker Heights, Ohio, 1971, 151–158.

[22] Ido Dagan, Oren Glickman, and Bernardo Magnini, The PASCAL recognising textual entailment challenge, *Machine Learning Challenges*, Berlin: Springer Verlag, 2006.

[23] Charles Darwin. *On the Origin of Species*. London: John Murray, 1859.

[24] Ernest Davis. *Linear Algebra and Probability for Computer Science Applications*. Boca Raton, FL: CRC Press, 2012.

[25] Richard Dawkins. *The Selfish Gene*. New York: Oxford University Press, 1976.

[26] Terrence Deacon. *The Symbolic Species: The Co-Evolution of Language and the Brain*. New York: W. W. Norton & Co., 1997.

[27] James Delgrande. A framework for logics of explicit belief. *Computational Intelligence* 11 (1985): 47–88.

[28] Daniel Dennett. *Brainstorms: Philosophical Essays on Mind and Psychology*. Cambridge, MA: MIT Press, 1981.

[29] Daniel Dennett. *The Intentional Stance*. Cambridge, MA: MIT Press, 1987.

[30] Daniel Dennett. Précis of *The Intentional Stance*. *Behavioral and Brain Sciences* 11 (1988): 495–505.

[31] Daniel Dennett. *Kinds of Minds: Towards an Understanding of Consciousness*. London: Weidenfeld & Nicolson, 1996.

[32] Zoltan Dienes and Josef Perner. A theory of implicit and explicit knowledge. *Behavioral and Brain Sciences* 22 (1999): 735–755.

[33] Norman Doidge. *The Brain That Changes Itself: Stories of Personal Triumph from the Frontiers of Brain Science*. London: Penguin Books, 2007.

[34] Merlin Donald. Origins of the modern mind: three stages of the evolution of culture and cognition. *Behavioral and Brain Sciences* 16 (1993): 737–791.

[35] Robin Dunbar. *Grooming, Gossip and the Evolution of Language*. London: Faber and Faber, 1996.

[36] Hubert Dreyfus. *What Computers Still Can't Do: A Critique of Artificial Reason*. Cambridge, MA: MIT Press, 1992.

[37] Hubert Dreyfus and Stuart Dreyfus. *Mind over Machine: The Power of Human Intuition and Expertise in the Era of the Computer*. New York: Free Press, 1986.

[38] Paul Edwards, ed. *The Encyclopedia of Philosophy*. New York: Macmillan, 1967.

[39] Oren Etzioni, Michele Banko, Stephen Soderland, and Daniel Weld. Open information extraction from the web. *Communications of the ACM* 51 (2008): 68–74.

[40] Ronald Fagin and Joseph Halpern. Belief, awareness and limited reasoning. *Artificial Intelligence* 34 (1987): 39–76.

[41] Ronald Fagin, Joseph Halpern, Yoram Moses, and Moshe Vardi. *Reasoning About Knowledge*. Cambridge, MA: MIT Press, 1995.

[42] Jerry Fodor. *The Mind Doesn't Work That Way*. Cambridge, MA: MIT Press, 2000.

[43] Lance Fortnoy. The status of the P versus NP problem. *Communications of the ACM* 52 (2009): 78–86.

[44] Michael Genesereth and Nils Nilsson. *Logical Foundations of Artificial Intelligence*. Los Altos: Morgan Kaufmann, 1987.

[45] Edmund Gettier. Is justified true belief knowledge? *Analysis* 23 (1963): 121–123.

[46] Tatiana Giraud, Jes Pedersen, and Laurent Keller. Evolution of supercolonies: The Argentine ants of southern Europe. *Proceedings of the National Academy of Science* 99 (2002): 6075–6079.

[47] Carla Gomes, Henry Kautz, Ashish Sabharwal, and Bart Selman. Satisfiability solvers. In *Handbook of knowledge representation: Foundations of Artificial Intelligence*, ed. F. van Harmelen, V. Lifschitz, B. Porter. Amsterdam: Elsevier, 2008, 89–134.

[48] Carla Gomes, Bart Selman, Nuno Crato, and Henry Kautz. Heavy-tailed phenomena in satisfiability and constraint satisfaction problems. *Journal of Automated Reasoning* 24 (2000): 67–100.

[49] Malcolm Gladwell. *Blink: The Power of Thinking without Thinking.* New York: Little, Brown and Co., 2005.

[50] Phillips Griffiths, ed. *Knowledge and Belief.* London: Oxford University Press, 1967.

[51] Armin Haken. The intractability of resolution. *Theoretical Computer Science* 39 (1985): 297–308.

[52] David Harel. *Algorithmics: The Spirit of Computing.* Reading, MA, Addison-Wesley, 1987.

[53] Stevan Harnad. The symbol grounding problem. *Physica D* 42 (1990): 335–346.

[54] John Haugeland. *Artificial Intelligence: The Very Idea.* Cambridge, MA: MIT Press, 1985.

[55] Samuel Hayakawa and Alan Hayakawa. *Language in Thought and Action.* New York: Harcourt Brace Jovanovich, 1991.

[56] Frederick Hayes-Roth, Donald Waterman, and Douglas Lenat. *Building Expert Systems.* Reading, MA: Addison-Wesley, 1983.

[57] Jaakko Hintikka. *Knowledge and Belief.* Ithaca, NY: Cornell University Press, 1962.

[58] Geoff Hinton, Jay McClelland, and David Rumelhart. Distributed representations. In *Parallel Distributed Processing*, ed. D. Rumelhart and J. McClelland. Cambridge, MA: MIT Press, 1986, 77–109.

[59] Nicholas Humphrey. The social function of intellect. In *Growing Points in Ethology*, ed. P. Bateson and R. Hinde. Cambridge, UK: Cambridge University Press, 1976, 303–317.

[60] David Israel. The role of logic in knowledge representation. *IEEE Computer* 16 (1983): 37–42.

[61] Peter Jackson. *Introduction to Expert Systems*. Reading, MA: Addison-Wesley, 1990.

[62] Philip Johnson-Laird, Sangeet Khemlani, and Geoffrey Goodwin. Logic, probability, and human reasoning. *Trends in Cognitive Science* 19 (2015): 201–214.

[63] Daniel Kahneman. *Thinking, Fast and Slow*. New York: Farrar, Straus and Giroux, 2011.

[64] Christine Kenneally. *The First Word: The Search for the Origins of Language*. New York: Penguin Books, 2007.

[65] Raymond Kurzweil. *The Singularity Is Near: When Humans Transcend Biology*. London: Viking Penguin, 2005.

[66] Quoc Le, Marc-Aurelio Ranzato, Rajat Monga, Matthieu Devin, Kai Chen, Greg Corrado, Jeffrey Dean, and Andrew Ng. Building high-level features using large scale unsupervised learning. In *Proceedings of ICML 2012: The 29th International Conference on Machine Learning*, Edinburgh, Scotland, June 2012, 81–88.

[67] Douglas Lenat and Ramanathan Guha. *Building Large Knowledge-Based Systems: Representation and Inference in the Cyc Project*. Boston: Addison-Wesley, 1990.

[68] Hector J. Levesque. A logic of implicit and explicit belief. In *Proceedings of AAAI-84: The Fourth National Conference on Artificial Intelligence*, August 1984, 198–202.

[69] Hector J. Levesque. Logic and the complexity of reasoning. *Journal of Philosophical Logic* 17 (1988): 355–389.

[70] Hector J. Levesque. Is it enough to get the behavior right? In *Proceedings of IJCAI-09: The 21st International Joint Conference on Artificial Intelligence*, Pasadena, California, August 2009, 1439–1444.

[71] Hector J. Levesque. The Winograd Schema Challenge. In *Proceedings of Commonsense-11: The Tenth International Symposium on Logical Formalizations of Commonsense Reasoning*, Palo Alto, March 2011, 53–58.

[72] Hector J. Levesque. *Thinking as Computation: A First Course.* Cambridge, MA: MIT Press, 2012.

[73] Hector J. Levesque. On our best behavior. *Artificial Intelligence* 212 (2014): 27–35.

[74] Hector J. Levesque, Ernest Davis, and Leora Morgenstern. The Winograd Schema Challenge. In *Proceedings of KR 2012: The Thirteenth International Conference on Principles of Knowledge Representation and Reasoning*, Rome, June 2012, 552–561.

[75] Vladimir Lifschitz, ed. *Formalizing Common Sense: Papers by John McCarthy.* Exeter, UK: Intellect, 1998.

[76] Marc Lipson and Seymour Lipschutz. *Schaum's outline of theory and problems of linear algebra.* New York: McGraw-Hill, 2001.

[77] Alan Mackworth. On seeing robots. In *Computer Vision: Systems, Theory and Applications*, ed. A. Basu and X. Li. Singapore: World Scientific Press, 1993, 1–13.

[78] John Macy, ed. *The Story of My Life.* New York: Doubleday, Page & Co., 1905.

[79] John McCarthy. Programs with common sense. In *Proceedings of Symposium on the Mechanization of Thought Processes.* National Physical Laboratory, Teddington, England, 1958, 77–84.

[80] Pamela McCorduck. *Machines Who Think.* 25th anniversary edition. Natick, MA: A K Peters, 2004.

[81] Marvin Minsky, ed. *Semantic Information Processing.* Cambridge, MA: MIT Press, 1968.

[82] Marvin Minsky. A framework for representing knowledge. In *Readings in Knowledge Representation*, ed. R. Brachman and H. Levesque. San Francisco: Morgan Kaufmann, 1985, 245–262.

[83] Marvin Minsky. *The Society of Mind.* New York: Simon & Schuster, 1986.

[84] Marvin Minsky. Logical versus analogical or symbolic versus connectionist or neat versus scruffy. *AI Magazine* 12 (1991): 34–51.

[85] Robert Moore. The role of logic in knowledge representation and commonsense reasoning. In *Proceedings of AAAI-82: The Second National Conference on Artificial Intelligence*, August 1982, 428–433.

[86] Kevin Murphy. *Machine Learning: A Probabilistic Perspective.* Cambridge, MA: MIT Press, 2012.

[87] Monty Newborn and Monroe Newborn. *Deep Blue: An Artificial Intelligence Milestone.* Berlin: Springer Science & Business Media, 2003.

[88] Allen Newell and Herbert Simon. Computer Science as empirical inquiry: Symbols and search. *Communications of the ACM* 19 (1976): 113–126.

[89] Nils Nilsson. *The Quest for Artificial Intelligence: A History of Ideas and Achievements.* Cambridge, UK: Cambridge University Press, 2009.

[90] Mike Oaksford and Nick Chater. *Bayesian Rationality: The Probabilistic Approach to Human Reasoning.* New York: Oxford University Press, 2007.

[91] Taketoshi Ono, Ryoi Tamura, and Kiyomi Nakamura. The hippocampus and space: are there "place neurons" in the monkey hippocampus? *Hippocampus* 1 (1991): 253–257.

[92] Don Perlis, Khemdut Purang, and Carl Andersen. Conversational adequacy: mistakes are the essence. *International Journal of Human-Computer Studies*, 48 (1998): 553–575.

[93] Steven Pinker. *The Language Instinct.* New York: Harper Perennial Modern Classics, 1994.

[94] Steven Pinker. *How the Mind Works.* New York: W. W. Norton, 1999.

[95] Steven Pinker. *The Blank Slate.* New York: Viking, 2002.

[96] David Poole, Alan Mackworth, and Randy Goebel. *Computational Intelligence: A Logical Approach.* New York: Oxford University Press, 1998.

[97] Zenon Pylyshyn. *Computation and Cognition: Toward a Foundation for Cognitive Science.* Cambridge, MA: MIT Press, 1984.

[98] Willard Quine. Quantifiers and propositional attitudes. *Journal of Philosophy* 53 (1956): 177–187.

[99] William Rapaport. How Helen Keller used syntactic semantics to escape from a Chinese Room. *Minds & Machines* 16 (2006): 381–436.

[100] Raymond Reiter. On closed world data bases. In *Logic and Data-bases*, ed. H. Gallaire and J. Minker. New York: Plenum Press, 1987, 55–76.

[101] Matt Ridley. *The Red Queen: Sex and the Evolution of Human Nature.* London: Penguin Books, 1993.

[102] John Robinson. A machine-oriented logic based on the resolution principle. *Journal of the ACM* 12 (1965): 23–41.

[103] Philip Ross. The expert mind. *Scientific American* 295 (2006): 64–71.

[104] Michael Roth and Leon Galis. *Knowing: Essays in the Analysis of Knowledge.* Lanham, MD: University Press of America, 1984.

[105] Stuart Russell and Peter Norvig. *Artificial Intelligence: A Modern Approach.* Upper Saddle River, NJ: Pearson Education, 2010.

[106] John Searle. Minds, brains and programs. *Behavioral and Brain Sciences* 3 (1980): 417–424.

[107] Murray Shanahan. *The Technological Singularity.* Cambridge, MA: MIT Press, 2015.

[108] Stuart Shieber, ed. *The Turing Test: Verbal Behavior as the Hallmark of Intelligence.* Cambridge, MA: MIT Press, 2004.

[109] Robert Shumaker, Kristina Walkup, and Benjamin Beck. *Animal Tool Behavior: The Use and Manufacture of Tools by Animals.* Baltimore, MD: Johns Hopkins University Press, 2011.

[110] Herbert Simon and William Chase. Skill in chess. *American Scientist* 61 (1973): 394–403.

[111] Anne Sinclair, Robert Jarvella, and Willem Levelt, eds. *The Child's Conception of Language*. Berlin: Springer-Verlag, 1978.

[112] Brian Cantwell Smith. *Reflection and Semantics in a Procedural Language*. Cambridge, MA: PhD thesis, Massachusetts Institute of Technology, 1982.

[113] Paul Smolensky. Connectionist AI, symbolic AI, and the brain. *Artificial Intelligence Review* 1 (1987): 95–109.

[114] Paul Smolensky. Analysis of distributed representation of constituent structure in connectionist systems. In *Proceedings of NIPS-87: Neural Information Processing Systems*. Denver, Colorado, November 1988, 730–739.

[115] Larry Squire. Declarative and nondeclarative memory: Multiple brain systems supporting learning and memory. *Journal of Cognitive Neuroscience* 4 (1992): 232–246.

[116] David Stork, ed. *HAL's Legacy: 2001's Computer as Dream and Reality*. Cambridge, MA: MIT Press, 1997.

[117] Nassim Taleb. *The Black Swan: The Impact of the Highly Improbable*. New York: Random House, 2007.

[118] Alan Turing. On computable numbers, with an application to the *Entscheidungsproblem*. *Proceedings of the London Mathematical Society* 42 (1937): 230–265.

[119] Alan Turing. Computing machinery and intelligence. *Mind* 59 (1950): 433–460.

[120] Herbert Turnbull. *The Correspondence of Isaac Newton: 1661–1675*. Volume 1. London: Cambridge University Press, 1959.

[121] Bonnie Webber and Nils Nilsson, eds. *Readings in Artificial Intelligence*. Los Altos, CA: Morgan Kaufmann, 1981.

[122] Joseph Weizenbaum. ELIZA. *Communications of the ACM* 9 (1966): 36–45.

[123] Joseph Weizenbaum. *Computer Power and Human Reason: From Judgment to Calculation.* New York: W. H. Freeman & Co., 1976.

[124] Jeanette Wing. Computational thinking. *Communications of the ACM* 49 (2006): 33–45.

[125] Terry Winograd. *Understanding Natural Language.* New York: Academic Press, 1972.

[126] Terry Winograd. Frame representations and the declarative/ procedural controversy. In *Representation and Understanding: Studies in Cognitive Science,* ed. D. Bobrow and A. Collins. New York: Academic Press, 1975, 185–210.

Index

Printed in the United States
by Baker & Taylor Publisher Services